U0422408

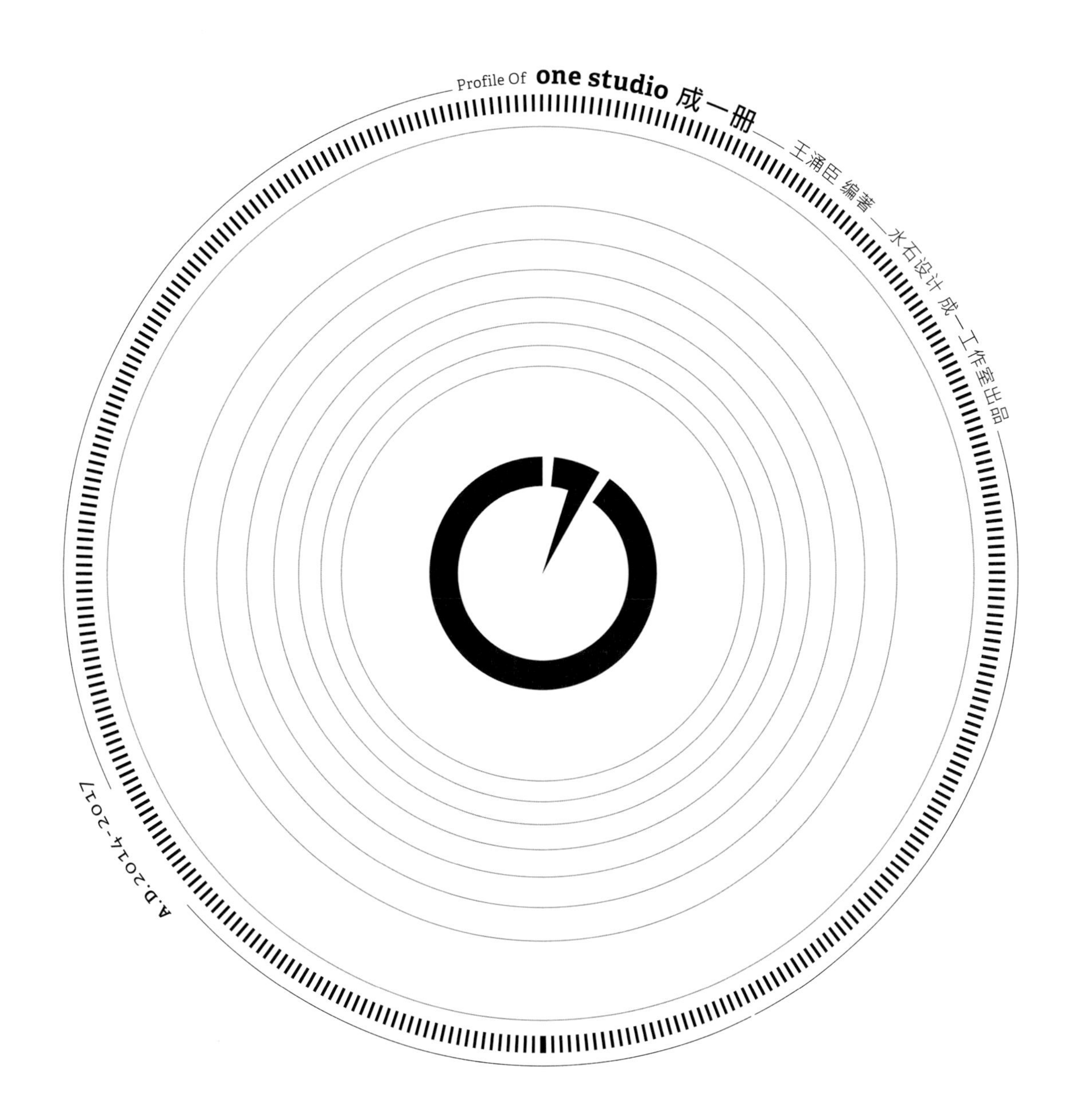
Profile Of one studio 成一册
王涌臣 编著
水石设计 成一工作室出品
A.D.2014-2017

水石设计
WWW.SHUISHI.COM

同濟大學出版社
TONGJI UNIVERSITY PRESS

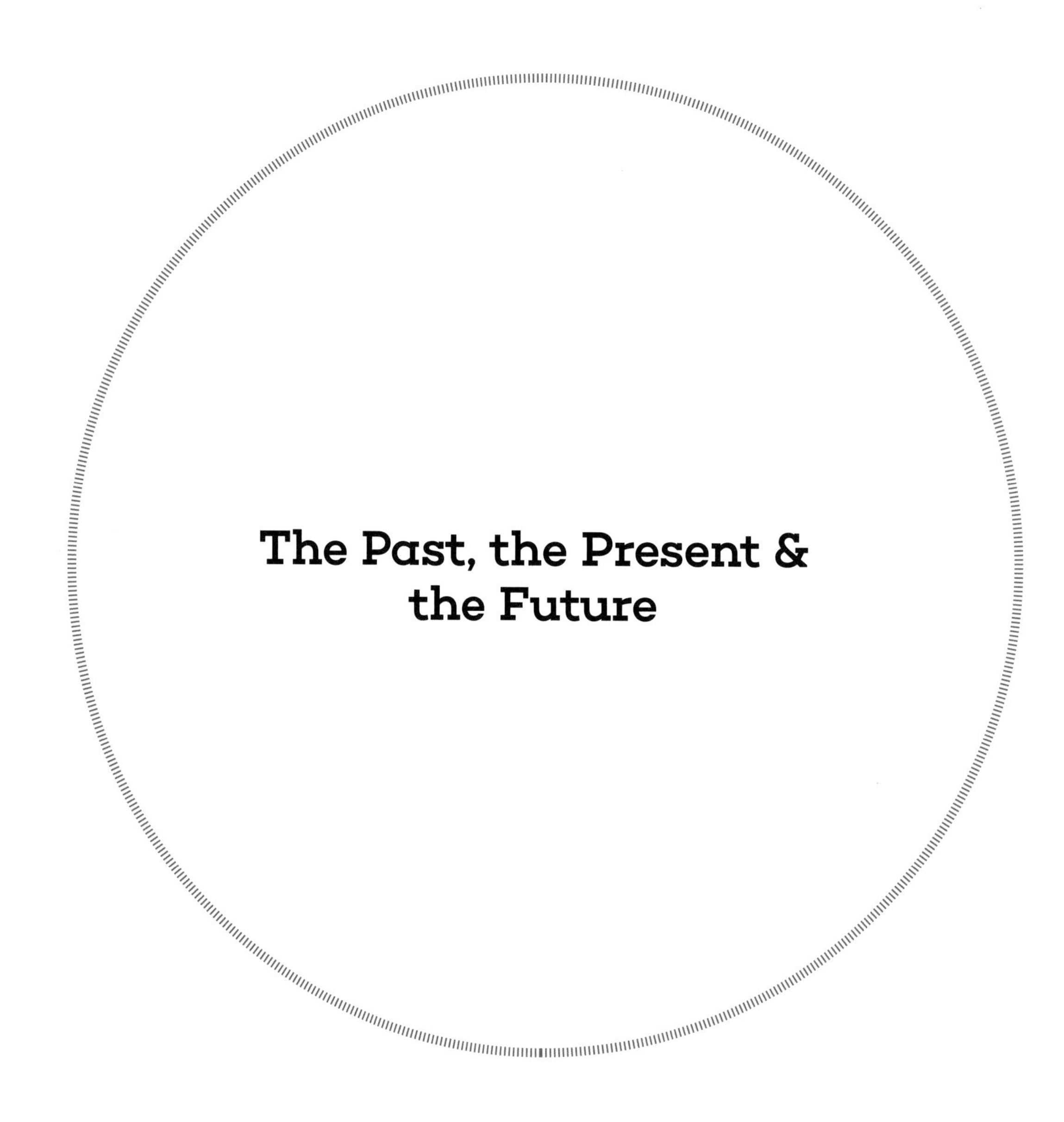

The Past, the Present & the Future

艺术、创造力和幻想的世界一直非常神秘，

从未有人揭示过（能够让人众所周知）一个创意是如何产生的，

一件作品是如何被创作出来的……然而，我认为人们希望了解，

因此我应该试图去解释。

布鲁诺·穆纳里 《幻想》拉泰尔扎 1977 年

one studio

成一工作室

Insightful Architecture

An Insightful Design Studio
Since a.d. 2017, by
W&R Group, Shanghai

188 DESIGN
No.188 Guyi Road, Xuhui District,
Shanghai, 200235
+86-021-54679918

目录
Catalogue

谦虚的设计
Modest Design

锚入风土
Anchoring into the Customs

1 2 3 4 5

序/ 水石的选择

项目资源越来越集聚，市场越来越成熟，业主越来越专业；高水平设计人才在各方都非常稀缺，情怀、能力、执着兼备者尤其难见。这就是我们当前面临的形势，未来属于“大而强”和“小而美”的机构。水石很幸运，遇到一个持续变化发展的时代，依据需求和自身能力特点，水石选择了一条规模化、综合化的发展途径。

设计的本质是“智慧创造价值”，当下市场设计的需求正呈现出多元化趋势。设计既有技术性、综合性的一面；也有个性化、差异性的需求；设计服务如何兼顾，水石一直在试图通过人才培育与储备来夯实基础。相比规模化、综合化，引导与培育具有设计高度、锐度，以及专业、专项能力的人才与团队更为急迫，这关系到水石未来的竞争优势，以及持续稳步发展的宽度与高度空间。

—— 邓刚 董事长、创始合伙人 [1]

成一工作室的成立是水石发展过程中的一次重要尝试。水石希望涌臣带队，将富有创造力又年富力强的设计师，组成先锋团队，作为敢闯敢做的年轻设计师充分发展的空间。成立之初，经管理层和团队多次讨论，成一的发展方向定位为城市更新、产业地产、特色公建领域。虽然水石的优势设计能力仍在居住地产模块，但随着居住地产黄金时代的逐步褪色，水石迫切需要在城市更新、产业地产、特色公建等更多领域发展，开拓并建立新的核心竞争力。希望成一工作室能在其中担当更加重要的角色。

—— 倪量 董事、总裁、创始合伙人 [2]

建筑设计本身就是一门和形式美息息相关的学科和行业，而成果表达的效果本身就是体现设计师水准的重要方面。最优秀的设计团队往往在细节上有格外的追求，讲究图纸的精美、逻辑的精彩、作品集的品质等。成一工作室正向其看齐，用他们自己的语言向读者传达某种特质。成一从项目类型上有主题产业园、超高层办公、五星酒店及城市更新设计。客户既有知名开发商也有大型国企，既有民营企业也有政府机构。在如此大的跨度中，每个领域他们都做出了很出彩的内容，实属不易。

然而，通往这些优秀作品，艰辛和波折都是必经之路。纵然坎坷而漫长，但优秀的建筑师就是要把一个个项目熬到完美的建成。期待在下一部作品集里，能看到成一更多的建成作品和优秀的设计。

—— 严志 董事、执行总裁、创始合伙人 [3]

成一工作室前身为商业公建部的 C3 团队，在涌臣的带领下取得了非常耀眼的成绩。这次以独立工作室的角色再一次总结最近两年的完整项目经历，确实非常有内容！新颖的视角，缜密的逻辑，同时极高的颜值，都是这本书册给我的鲜明印象。相信所有看到这本册子的读者都会像我一样，感受到这些设计师对待自己职业的热爱与诚意，这也是涌臣以“成一”命名工作室最重要的用意吧。

非常荣幸也非常骄傲，能与“成一”这样的团队共事。众所周知，设计行业在近些年随着地产行业的剧烈变化同样剧变着，正被描述成全社会最疲惫的行业之一，设计师的各种被催残的故事，也频频在行业内甚至社会上流传，一时间建筑师成为最苦最累最没有品质生活的职业代名词！这几乎直接导致了近两三年持续有大量建筑师改行或离开大城市，成为现象级的行业趋势。但我感觉这很正常，大浪淘沙的结果是留下好沙，浮土和杂质被冲刷掉了。建筑师是终身职业，成长期长，也是最古老传统的职业，是跟着人类城市历史不断积累荣耀的职业，是最重要的、传递人类精神意志的工作之一，建筑师几乎是永久地改变着人类生活环境的面貌，这份职业的尊严需要热爱和诚意坚守！在‘成一’团队的这本书册里，我们看到了这样的职业尊严，我觉得所有的读者也会情不自禁地鼓掌喝彩，这是我们职业信心的佐证和来源。

—— **王煊 董事、总建筑师、创始合伙人** [4]

2017 年水石有很多变化，搬入新的 Design188 共享设计空间，业务保持快速增长，同时设计师队伍也在不断扩大，‘专业化’‘全程化’已经成为公司发展的共识。水石纵向形成全流程各专业设计部门，横向并置为数个设计部门并形成规模。在这样的市场化、规模化框架之下，如何激励设计师投入到更多元的设计领域中去，并保持创作的活力，是需要长期研究的问题。

从加入水石起，涌臣团队的设计风格给人的印象非常统一，很大程度上我想是因为视觉表达的高度一致。这也让人思考图示语言对设计的影响有多大。这样的设计方法，是结合水石的策划经验与图形生成（diagram）的手法。例如光明冷库改造项目中，空间的图形生成就是最主要的手法，同时这种几何秩序的建立又源于项目本身的开发策略。

“成一”的另一个特点是对自我的总结，并积极分享信息，深度参与水石的技术平台搭建以及培训计划，一方面影响着平台上其他团队的设计师，另一方面也在这个过程中总结和提高技术能力。值得一提的是，他们拥有特强的设计策划能力，从几年前宝山煤气厂的更新规划到最近的物流园区设计，应该说一脉相承着核心的产业园区运营及设计经验。除此以外，后期项目的落地实施也有很多可圈可点的技术积累，从宝山绿地创新中心到南京瑞安珠江路立面改造项目。期待“成一”在未来有更大的发展。

—— **沈禾 董事、设计总监、创始合伙人** [5]

照片来源：2017 年秋季 水石设计 - 技术研讨会“讲武堂”，佘山，上海

Preface/ The Selection of W&R

The project resources are increasingly concentrated.The market is increasingly mature. The proprietors are increasingly professional.High-level designers are very scarce on all sides. It's especially hard to find people who have feelings,abilities and persistence. This is the situation we are facing currently. The future belongs to the "big and strong" and "small and beautiful"institutions. W&R is in the era of continuous change and development, which is quite lucky. Based on requirements and its individual characteristics and abilities, W&R selects a large-scale and comprehensive development approach.

The essence of design is "wisdom creating value". Nowadays, the demand for market design is presenting a trend of diversification. On one hand, design is technical and comprehensive. On the other hand, it needs to be personalized and differentiated. How to balance design and service?

W&R is trying to lay a solid foundation by cultivating and reserving talents. Compared with scale and totalization, it's more urgent to guide and cultivate talents and teams that have the height and acuity of design as well as professional and special abilities which are related to the competitive advantage of the W&R in the future as well as the width and height of continuous and stable development.

-Chairman Deng Gang, Founding Partner[1]

The establishment of One Studio is an important attempt in the process of the development of W&R. W&R wants Yongchen to lead a pioneer team where creative and energetic designers gather. The team will be a space for the full development of young designers. At the beginning of the establishment, the development direction of One Studio are the renewal of cities, industrial real estate and the public construction that is distinctive. W&R's ability of preponderant design is still living estate. However, as the golden age of living estate fades, there is an urgent need for W&R to develop in more areas, such as the renewal of cities, industrial real estate and the public construction in roder to exploit and establish new core competitiveness. I hope One Studio will play a more significant role.

-Niling Director, President, Founding Partner[2]

Architectural design itself is a discipline and profession that is closely linked with formal beauty. The effect of achievement itself is an important aspect that reflects the level of designers. The best team are incomparably focus on details, such as the delicacy of diagram paper, wonderful logic, the quality of sample reels and so on. One Studio is keeping up with it and convey a certain kind of trait by their own language. The type of 's project are theme industrial park, ultra-high-rise offices, five-star hotels and the design of urban renewal. Clients are well-know developers, large state-owned enterprises, private enterprises and governmental agencies. In such a wide span, it is not easy for them to do well in every field. However, the only sure way to these excellent works are hardship and setback. Although the process are probably rough and lengthy, outstanding architects didn't give up projects until they were finished perfectly. I hope see more completed projects and good designs.

-Yanzhi, Chairman, CEO, Founding Partner[3]

The predecessor of One Studio was C3 team of commercial public department that achieved good results under the leadership of Yongchen. This time, as an independent studio, One Studio summarized complete which is indeed substantial. The novel visual angle, meticulous logic and the attractiveness of cover made a strong impression on me. I believe that all readers of this booklet will feel these designers'passion and sincerity for their occupation. This is why the studio is named after One Studio by Yongchen.
I am honored and proud to work with a team like One Studio. As we all know, with the dramatic changes of real estate industry, the design industry that is described as one of the most exhausting industry in the whole society upheavals similarly. The stories of shattered designers frequently spread in this industry, even society. Architects became a synonym for the hardest and the most tired profession in which there is no high quality of life. As a result, a large number of architects have continually changed positions or leaved big cities in the last two or three years, which becomes a phenomenal industry trend. But I think it is normal. Good sand is left and topsoil and impurity are swept away after ebb tide. The profession as an architect is a lifelong career whose growth stage is long. It is also the oldest and the most traditional career that gather glory continually with the history of mankind. It is one of the most important careers that convey people's mentality. Architects have been nearly changing people's living environment timelessly. The dignity of this profession should be loved and kept sincerely. In the book of One Studio team, we feel this kind of professional dignity. I think all readers can not help applauding and cheering, which is the evidence and source of our confidence of the profession.

-Wangyu, Director, Chief Architect, Founding partner[4]

W&R has a lot of changes in 2017. It moved into the new Design 188 that is a space of shared design. The business keeps growing quickly and designer teams continuously extend. "Professionalization" and "integration" has become the consensus of the development of the company. W&R form the design departments of each profession in the entire process lengthways and concatenate several design departments that form scale breadthways. Under such a market oriented and large-scaled framework, the question how to inspire designers to devote themselves to the field of more diversified design and maintain the energy of creation need to be studied for a long time.
From the beginning of joining W&R, the design style of Yongchen team makes an unified impression on people. In my opinion, because the visual expression is highly consistent to a great extent, which makes people think of the great influence of graphic language. The way of design combine the planing experience of W&R with graph-generation. For example, in the project of reconstruction of Bright Shenhong Cold Storage, the major way is the graph-generation of space. Meanwhile, the establishment of geometric order comes from the exploitation strategy of the project itself.
Other characteristics of One Studio are self-summary, sharing information actively and deeply engaging in the establishment of technology platform and training program of W&R, which affects designers of other teams of the platform on the one hand and summarize and improve technical skills in the process on the other hand. It is worth mentioning that they have the advantaged ability of designing and engineering. From the programme of the renewal of Baoshan Gasworks a few years ago to the design of logistics park recently, it is in line with the core experience of operation and design of industrial park. In addition, there are much significant technology accumulation in the process of implementation of later projects, such as the project of Greenland Innovation Center and the facade renovation in Zhujing road, Nanjing. I look forward to a greater development of One Studio in the future.

- Shenghe, Director, Design Director, Founding Partner [5]

The source of Photo: The autumn of 2017, The Design of W&R Technology Seminar "Jiangwutang", Sheshan, Shanghai

自序 / 请让我把话说完

这是个跑马般的时代，每一个人都在不停地奔跑，超过三个逗号的语句都被视作冗余，更多话语被淹没在不耐烦的臆测中。我们很难有机会与人们交流，很难让思维中忽明忽暗的火花得到充分燃烧，话说明白已很难得，把话说完更是奢侈。因此在“成一”成立一周年这个特殊的时刻，我想通过作品集的形式，作为表达，同时也系统性地自我反思与总结。

在本书的编纂中，回顾了近几年的作品，我们不甘心让这些历程就此黯淡。一路奋行，我们完成的除了这 20 余件作品，更多的是经历了憧憬、质疑、争议、决策后不断架构的鲜活思维。在这 200 多页，请让我把话说完，留给 20 年后的自己，也留给正在浏览的你。

做漂亮的方案

设计是探索性的工作，很多时候在向人们展示设计前，他并不很确定自己究竟想要什么。这时候第一印象至关重要。所以，我们一直坚持做漂亮的文本，且不说设计本身是否能受到青睐，第一眼的体验必须是舒适的、美好的。我们决不允许因平庸的视觉体验而亏欠了设计的精华。

方案本身的美感，图纸的排版风格和细节，都关乎着客户的视觉体验，也体现我们的态度和诚意。毕竟，人是视觉动物，而我们也身在一个在乎“颜值”的世界。

用逻辑去说话

我们一直努力让设计恰如其分地适合于情景，尽可能摒弃刻意强调“我建议如此”的个人特征，根据实际情况来让设计契合各方面的约束条件。

这意味着，在各种复杂的条件下，我们需要判断形势，权衡利弊，揣测人心，给出多种解决办法，还要会随机应变。而这一切的基础就是前期大量的数据支持和逻辑分析。地块特征、环境优劣势、城市贡献、人文关怀、使用者的真实诉求等这一切的考量，都在引导我们 从纷繁的背景中抽丝剥茧，以最贴切的设计反映最本质的诉求，让创意在合情合理中出乎意料。

“成一”是个集体

建筑师天生是合作者，尤其在碎片化工作与社会协同合作的当代。设计需要交流、反思，需要在辩解中明确边界，需要在碰撞中摸索灵感的组合，需要在个性发挥中推进团队，需要在互补长短中共同成长。“成一”就是这样的一支团队，我们相信，设计中集群智慧的人文价值与设计师的精神品格是密不可分的。

“成一”这个名字源于我们一个初衷——成就每一个项目。制作这本《成一册》也就是希望将我们的过去装订成一册，一起展望未来。如同“成一”标识所示，这个团队还很年轻，正在一点钟刚刚出发，也正因为年轻，我们无所忌，尽己所能的聚合智慧完成每一件作品。同时，在这里成就的不仅仅是一个个项目，也是这支团队里每一名成员。

不急着“成熟”

作为坚信创意会为现实带来梦想的设计师，我们拒绝简单的“成熟”，拒绝将眼前的事实与现实画等号，尤其在这个并不成熟的市场环境下——碎片化的项目、反复无常的节奏、开发价值为主体导向、充满风险与未知的市场，在泡面式操作中趋于套路的“匠心”。设计师需要理解现实，要客观，但如果将自我的这份热忱都删除后，我们所观察到的就只剩客观，而没有真正意义上的现实了。我们坚信设计的力量，我们坚信设计师的生命力，所以我们不厌其烦反复还原设计本质进行思考，用逻辑与想象为现实添彩，用诚意与坚持走向我们能够接受的“成熟”。

“成一”是为了一个共同的设计梦成就于每一天、每一个项目的团队，作为“水石人”，我们在自由中发展，在竞争中欣赏，在合作中凝聚。在面对未来的抉择时，一切的磨练与成长，最终会不知不觉地汇流成一股力量。

从创意到执行到最终呈现，从想到一个念头到做到一件作品，在所有材料的组合和处理中，以及与空间的呼应中，找到一种我们认为最好最准确的状态。绳锯木断，水滴石穿。

通过坚持与信念，抵达自己，我们不急着“成熟”。

王涌臣，成一工作室 主持建筑师

2017 年 12 月 于上海

Admission / Please Let Me Finish

This is an era where everyone is on the move like a running house. Sentences of more than three commas are considered redundant. More words are submerged in the impatient speculation. It is difficult for us to have the opportunity to communicate with people and to let the flickering spark of thoughts burn fully. It's not easy to say clearly and it's even luxurious to complete sentences. Therefore, in this special time of the first anniversary of the establishment of One Studio, I want express myself, introspect and summarize systematically by the form of sample reels. In the process of compiling the book, we review the works in recent years and are not willing to let them fade. Fighting all the way, we have experienced the vivid thinking that is constantly structured after longing, questioning, controversy and decision except for these about 20 works. In the more than two-hundred pages, please let me finish my words that are left to me 20 years later and you who are reading.

Do Elegant Programs

Design is an exploratory work. Usually, he is not sure what he really wants on earth before showing people the design. At this time, the first impression is crucial. Therefore, we always insist on making elegant texts. Regardless of whether the design itself can be favored, the first experience must be comfortable and nice. We are in no way to allow to owe the essence of design because of the visual experience that is mediocre.

The aesthetic feeling of design itself and typographic style and details of the graph paper are all related to the visual experience of customers and reflect our attitude and sincerity. After all, people are visual animals and we are in the world where people care about the outer attractiveness.

Speaking Logically

We have been working hard to make the design suitable for the situation rightly. We try to give up deliberately emphasizing the personal characteristics of "I recommend so" as far as possible and design the constraint conditions that meet various aspects according to the actual situation.

This means that we need to judge the situation, weigh the pros and cons, measure people's minds, give a variety of solutions, and adapt to it under all kinds of complicated conditions. The basis of them is a lot of data support and logic analysis in the early stage. The considerations of the characteristics of land parcel , environmental advantages and disadvantages, urban contribution, humanistic care and users' real appeals lead us to shed light on the complex backgrounds and to reflect the most essential demands by the most appropriate designs, making the creativity reasonably unexpected.

One Studio is a Collectivity

Architects are natural partners, especially in the contemporary era of fragmented work and social collaboration. Design requires communication and reflection. It requires a clear boundary in the process of defense, combination of inspirations in the collision, promotion of the team in the play of individuality and mutual development in complementation of each other. One studio is such a team. We believe that the humanistic value of cluster wisdom of design is inseparable from the designers' spiritual character.
The name of One studio stems from our original intention - accomplishing each project. The reason of making this book of One studio is the bookbinding of our past and to look into the future together. As the logo of One studio shows, this team is still very young and is just starting at one o'clock. So we have no taboos and do our best to assemble wisdom to complete each work. At the same time, the achievement here is not just a single project but also each member of this team.

Not in a Hurry to be "Mature"

As a designer who firmly believes that creativity will bring dreams to reality, we refuse to the simply "maturity" and to equate the facts with reality, especially under the situation of immature market - fragmented projects, repeatedly uncertain rhythm, the development of value-oriented, risky and unknown market, the ingenuity that tends to be routine in instant noodle operation. Designers need to understand the reality and be objective. But if we remove the hot pillow of ourselves, we can only observe objectively without the true reality. We firmly believe in the power of design and the vitality of designers. Therefore, we are willing to think by returning to the essence of design repeatedly. We add Color to reality with logic and imagination and move toward the "maturity" that we can accept with sincerity and persistence.
One studio is a team where there is a common design dream for each project everyday. As a member of W&R, we develop freely, appreciate things in the process of competition and unite in cooperation. In the face of future choices, all the hone and growth will eventually converge into one force unconsciously.
From the idea to the execution then to the final presentation, from thinking to accomplishing a work, in the combination and processing of all materialsand in the echo of space, we find a state that we believe is the best and most accurate. The wood is cut off by ropes and constant dripping wears away a stone.
We arrive at ourselves by persistence and faith. We are not in a hurry to be "mature".

Yongchen Wang, One studio, The Major Architect
2017.12 in Shanghai

WorkLife

2014

2017

兰州文化产业发展孵化中心
超高层
273 720m²
榆中县贾家庄美丽乡村
村落规划
30 600m²
兰州第一人民医院业务综合楼
医院 造型
45 682m²
兰州城关区文体医教综合
公建综合
570 847m²
兰州润通广场
超高层
229 700m²
成都龙湖时代天街策划
策划
651 094m²
郑州航空港区生物医药园
504 815 m²
南京珠江路未来城外
玄武 / 城市更新
37 900 m²
苏州金鸡湖 Indigo
酒店（五星）
75 201 m²
嘉善金地中德智能制
造及产业升级基地
920 600 m²
广西南宁世茂五象国际中心
超高层 综合体
325 263 m²
广州金地未来系列研发
广州实地集团禾丰 9# 地块
街区商业
108 640 m²
海南金泰湾盛泰澜度假酒店
酒店（五星）
115 967 m²
长春水生态文化园七净车间
南关区 / 城市更新
400 m²
青岛金地世家启动区
崂山区 / 住区
2 361 m²
上海国际能源创新中心
宝山区 / 城市更新
897 100 m²
上海绿地上海创新产业中心
宝山区 / 产业园
62 200 m²
上海光明申宏冷库改建项目
杨浦区 / 城市更新
78 750 m²
上海小绵羊总部产业园
嘉定区 / 城市更新
58 864 m²
上海海博西郊冷链物流园
青浦区 / 产业园
89 900 m²
上海赤峰路光明进修学院
杨浦区 / 城市更新
9 160 m²
上海蓝天游戏数字产业园
嘉定区 / 城市更新
76 705 m²
上海国际生态商务区 6 号地
松江区 / 酒店及公寓
123 291 m²

设计故事
The Stories of Design

理解客户品味 / 兰州文化产业发展孵化中心
Understanding Customer's Taste / Development and Incubation Center of Culture Industry in Lanzhou

纯粹的风格 / 兰州润通广场
Pure Style / Runtong Square in Lanzhou

解冻历史 · 再塑品牌 / 上海光明申宏冷库改造
Thawing History, Remolding Brands / The Reconstruction of Bright Shenhong Cold Storage in Shanghai

东方“船说”的故事 / 东方盛泰澜金泰湾度假酒店（五星）
The Story of "Ship" in the East / Dongfang Shengtailan Jintai Bay Resort Hotel (Five Star)

看起来不像，其实就是 / 兰州市第一人民医院综合楼
Not Look Like, It's Actually / The Comprehensive Business Building of the First People's Hospital of Lanzhou

是设计里的故事，还是故事里的设计？
设计需要故事。不仅要陈述故事，而且需要以表演的形式宣扬创意。令人愉悦的故事不仅要符合价值逻辑，更需要超越经验与思考，达至本能情感的驱动力。

在设计创意领域，总有一些有关故事的案例。1995 年，皮克斯计划发布《玩具总动员》。作为人类史上尝试的第一个电脑动画类电影，尽管当时电影界并不看好，普遍认为主流观众都不会关注，但随后却创造了电影票房的奇迹。在皮克斯之后的 16 部影片当中，8 部获得奥斯卡最佳动画片奖。

即使当时整个媒体都震惊了，但团队的成员们却似乎早已看到成功的结局，因为他们相信“自己能讲个好故事”。Ed Catmull 作为皮克斯的联合创始人之一，在他的畅销书 *Creativity, Inc.* 中写到关于皮克斯的创作过程：“如果你不能确保故事的准确性，所有你投入在艺术效果和视觉美感上的精力都白费。”

现在再也没有其他公司能达到这种入围率，说到底，还是通过故事这种联系来引人知悉、让人信任、引导消费。讲述一个好的故事，创建一种联系，终将提升作品的魅力，无论是通过电影、新闻或者建筑这种媒介。

1

Galerija
Gallery
ギャラリー

观九洞，素流分岩
"Nine holes" Scenery with Hills and Waters

张佾媛
Yiyuan，Zhang

镜片，麻纸，水墨、碳
Lenses，Hemp Paper, Ink, Carbon
原尺寸 Original Size 33mm × 33mm
2017

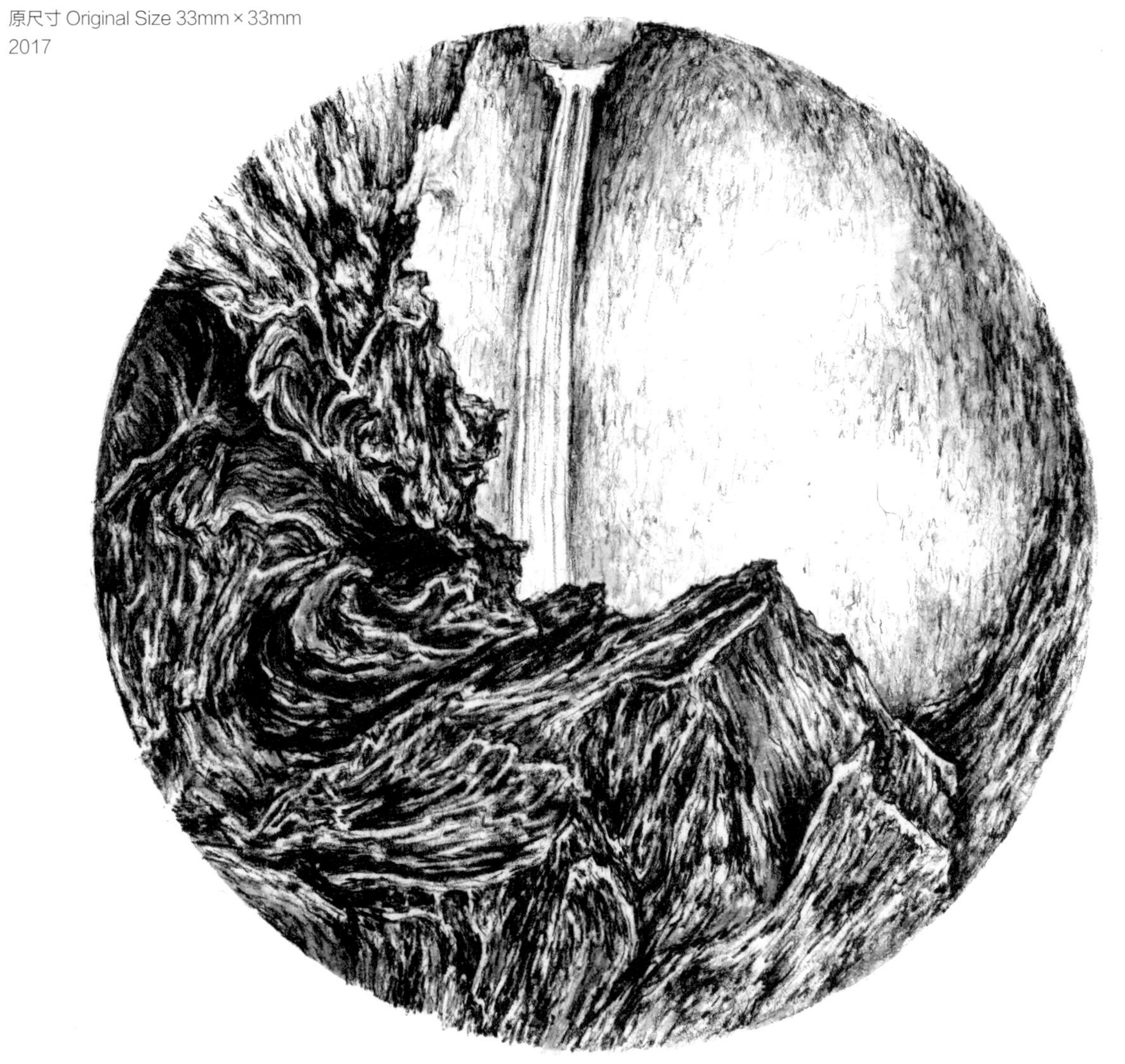

理解客户品位 / 兰州文化产业发展孵化中心
Understanding Customer's Taste / Development and Incubation Center of Culture Industry in Lanzhou

兰州

占地面积 ： 46 709 平方米

建筑面积 ： 273 720 平方米

建筑高度 ： 270 米

设计时间 ： 2014

Lanzhou

Floor Space : 46,709 m^2

Gross Floor Area : 273,720 m^2

Building Height : 270 m

The Time of design : 2014

传承正和情怀定制
Inheritance and Feelings Customized

拿到这个项目时，业主已组织过多家设计公司经过数轮探讨，上个团队的概念汇报时间更长达 5 小时（后来甚至闻听，曾有某个方案模型在汇报中途便被业主摔到楼下）。对项目特性和业主个性的重新审视，我们感觉是唯一能将设计继续开展下去的解题方式。

核心地段的地标尤其需要反映城市的文化本色。该项目是位于甘肃省政府和兰州市政府之间的 270 米超高层，距离黄河仅 200 米，近依白塔山，正面临主干道张掖路，地段优秀而敏感。为了符合政府形象和城市风貌的需求，设计需要符合区域庄严端正的风格，这是共识，但难点是，如何定制设计契合这位独特业主强烈的文化情怀与真实诉求。

了解并理解业主

业主正和集团是兰州的知名开发商，这是我们首次合作。慎重起见，在设计开始前，我们调研了其以往开发的所有项目、企业领导在媒体曾发表的观点，甚至还收集分析了正和会所的室内设计，我们尝试梳理出业主的特点，以进一步理解其核心关切的诉求。在充分调研后，我们认知到正和在其房地产开发中一直秉持品质及传承情怀，有深厚的中国传统文化底蕴，这也为后续设计奠定了文化、正气与雅致的重要设计基调。

设想不同寻常的组合

在如此敏感的区域、定位于超高层地标下的“古玩市场”、不允许造型使用大面积幕墙、需要传统文化的理解等条件及要求，为设计增加了不少难度。在随后的头脑风暴中，我们探讨了一个有趣的设想：在“博物馆”里卖古玩，以一座“光塔”流溢的金色强调“传承”“传统”“传世”。设计也由此展开：以气势、质感强调品质，以对称方正的形式为母题，探讨数字模数与企业图腾的置入，用黄金比例划分每段切割与收分。最终产生的方案，对传统交易市场和符合业主品位的超高层造型都是一次创新尝试。

When we got this project, the owners had organized several design companies to go through several rounds of discussions. The last team's concept was reported for up to five hours (Later we hear that there was a model of program that was thrown downstairs by owners in the process of reporting.) We think that the reexamination of projects' characteristics and owner's personality is the only way to solve the problem in order to continue the design.

The landmarks in the core area especially need to reflect the cultural qualities of the city. The project locate in the 270-meter super high-rise building between the provincial government of Gansu and the city government of Lanzhou. It is only 200 meters from the Yellow River, near Baita Mountain and is face the main road Zhangye Road. The site is excellent and sensitive. In order to meet the needs of the government's image and cityscape, the design needs to be in accord with the style of the region's solemnity, which is a consensus. However, the difficulty is how to customize the design to meet the strong cultural feelings and real demands of the unique owner.

Knowing and Understanding the Owners

The owner Zhenghe Group is a well-known developer in Lanzhou. This is our first cooperation. To be on the safe side, we investigated all the projects it had developed in the past and the opinions that corporate leaders had published in the media and even collected and analyzed the interior design of Zhenghe Club before the beginning of design. We tried to tease out the characteristics of the owners to further understand their appeals of core concerns. After full investigation, we recognize that Zhenghe that has deep Chinese traditional culture has always adhered to the quality and inherited feelings in its development of real estate, which also laid the important foundation of culture, righteousness and elegance for the later design.

Assume an Unusual Combination

Such a sensitive area, the "antique market" that locate in the super high-rise landmark, the inability to use large-scale curtain walls, the need for understanding of traditional culture and other conditions and requirements have added a lot of difficulties to the design. In the ensuing brainstorming, we explored an interesting idea: selling antiques in the "museum", emphasizing "transmitting", "tradition" and "passing down" with the gold of a "light tower" overflowing. Thereout, the design has also been carried out: emphasis on quality with momentum and texture, the motif in the form of symmetry, discussion of the placement of digital models and enterprise totems, and cuts and receipts of each segment with golden ratios. The resulting plan is an innovative attempt for the traditional trading market and super high-level modeling that meets the owners' tastes.

市中心、望黄河、观白塔
超高层、黄金比例切割、设计定制
建筑雕刻的是文化、彼此共鸣的是情怀

Downtown, Looking at the Yellow River
Appreciating the White Tower
Super High-rise, Cut in Golden Ratio, Customized Design
It is the culture that buildings sculpt. It is the feelings and they resonate with each other.

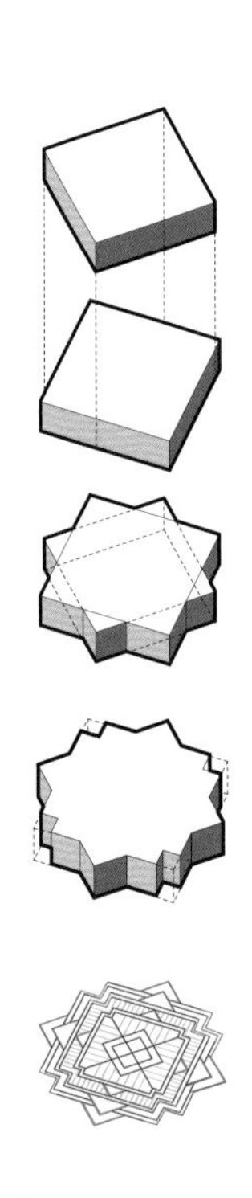

36°03′
34.1′ N
103°49′
39.3″ E

山峦起伏中，点亮城市的天际线
Lighting up the City's Skyline in the Undulating Hills

在定制设计中与业主共鸣

在设计中，我们将文化与方正贯彻于设计的每一步。塔楼的设计源于中国古建中密檐塔的轮廓造型与传统建筑窗花艺术结合，以套方为母题进行形式切割，用两个正方相叠，逐级退层，以黄金比例控制虚实对比，生成最终的形体。裙房古玩城的设计异于寻常市场风格，营造出博物馆的气质，彰显传统文化底蕴：九开间中轴对称，主体暖色石材，入口处立柱与两侧正方体块凸显气势、柱顶大红色企业标识图腾、8 片打磨后的汉白玉在灯光下尽演曼妙轻柔。此外，方案以九为基本模数，塔楼平面 45 米见方，裙房正面九跨开间，高 36 米，装饰构架也多以九为基本模数组合。究其缘由，也是取自中国传统文化中对九这个大数的尊崇。[1] 初次汇报后，业主评价“这是首次有设计团队的第一轮成果与我的想法契合度在 90% 以上”。回来后我们分析，这次成功相当一部分得益于设计方向的精准判断。

作为正和集团定制的地标，彰显当地经济文化腾飞的同时，也将作为正和企业精神的象征，赫赫屹立于黄河之巅。为了在 270 米高度上打造独一无二的天际线标志，我们特别设计了一条金色[2] 的光泉，自顶直贯入底，在城市与山峦的轮廓起伏中，如同黄河之水天上来。设计点亮的不仅是天际线，还有兰州人和正和人心中的文化信仰。文化与文物不同，文化始终鲜活地存在于创造中，建筑师除了价值创造，也有责任担当这份文化创造的义务。

目前项目仍在等待完成用地范围内多处拆迁工作，在项目具备条件后，成一将继续不断深化设计，将其打造成兰州中心区的地标性建筑群。

Resonating With Owners in Customized Design

We integrate culture and founder into every step. The tower design was derived from combination of the contour of Miyan Tower and art of window flower of traditional architecture. The form of square was used as the motif for cutting. Two squares were stacked one on top of the other and the layers were retreated step by step. We control contrast between fake and reality by golden ratio and final shape generates. Pimply antique city that is different from the style of usual markets creates atmosphere of the museum and highlights heritage of traditional cultures: It is axisymmetrical in nine rooms. The main body is made of warm-colored stone. The columns at the entrance and the cubes on both sides highlight momentum. The red logo totem of the corporation on the top of column and eight white polished marbles appear gracefully and softly under light. Besides, the program is based on the combination of number 9 as the basic module. The tower is 45 meters square. Nine-story podium whose height is 36 meters open front. Decorative framework also uses 9 because of the respect to it in Chinese traditional culture. After the initial report, owners commented that "It is the first time that the achievement of the first round of the design team meets my more than 90% goal." Then we analyzed that the success was considerably due to precision of design direction.

The landmark that is customized by Zhenghe Group highlights advancement of economy and culture of the region serve as a symbol of corporate spirit, standing majestically on the top of Yellow River. To create a unique mark of skyline at a height of 270 meters, we have specially designed a golden light spring that runs straight from top to bottom, which likes that Yellow River comes from heaven in the ups and downs of the contour of cities and mountains. It is not only skyline that is lightened up by design, but cultural beliefs in Lanzhou and Zhenghe. Unlike cultural relics, culture is always alive in creation. Architects also have responsibility to create cultures.
Nowadays, the project is waiting for completion of several demolition work within the scope of land using. Chengyi will continue to deepen its design and build it into an architectural complex that is the landmark in the center of Lanzhou.

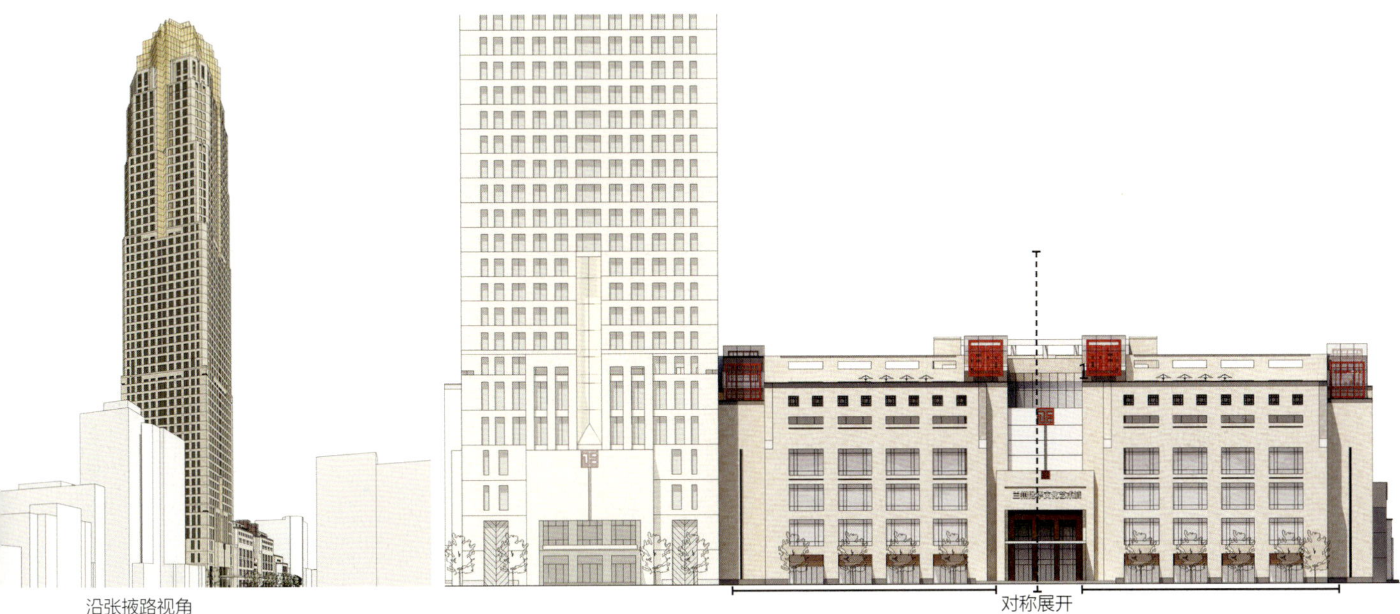

沿张掖路视角
The Zhangye Road Perspective

对称展开
Symmetric Expansion

注：

[1] 在中国古人的观念里，奇数为阳，偶数为阴，而奇数里最大的数字是"九"，故而古人对"九"这个数字特别重视，认为"九"为阳数之极，又可通译永恒。在中国传统文化中，"十"是满盈之数，物极必反，满则溢，极盛必衰，而"九"为"百尺竿头更进一步"，永远呈上升趋势，故"九"为至尊之数，为帝王所看中，皇宫建筑，多用"九"或"九"的倍数。故宫太和殿，五台山佛光寺大殿都是九间殿的代表。

[2] 兰州，始建于公元前 86 年。据记载，因初次在这里筑城时挖出金子，故取名金城。此外，还有一种说法是依据兰州城群山环抱，固若金汤，因此取"金城汤池"的典故，命名为金城，喻其坚固。

Note:

[1] In the conception of ancient Chinese, odd numbers were positive and even numbers were negative. The largest number in odd numbers was "nine". Therefore, the ancients who considered "nine" as the extreme number of positive numbers or eternity attached special importance to the number "nine". In the traditional Chinese culture, the "Ten" is a number of full surpluses. Things will develop in the opposite direction when they become extreme. Full water overflows. The extreme prosperity will certainly decline. But "Nine" will be "a further step", always showing an upward trend. So "Nine" that is valued by emperors is the supreme number. The buildings of the imperial palace usually use multiples of "nine" or "nine". The Taihe Temple of the Imperial Palace and the Buddhist Temple of Mount Wutai are representatives of the nine temples.

[2] Lanzhou was built in 86 BC. According to the records, Lanzhou was named Gold City because it was dug up for gold when it was first built here. In addition, there is another understanding that the city of Lanzhou is surrounded by mountains, as strong as iron, which is the literary quotation of "Iron Bastions". The name of Gold City means firmness.

汉白玉色对比暖色石材，入口突出
White Marble Color Contrast Warm Stone, Entrance Outstanding

博物馆就是最好的“古玩城”
The Museum is the Best Antique Market

庄严端正的风格，工作模型，摄于 2014 年 5 月
Solemn Style, Working Model 2014.5

纯粹的风格 / 兰州润通广场
Pure Style / Runtong Square in Lanzhou

兰州

占地面积	：10 825 平方米
建筑面积	：229 700 平方米
建筑高度	：270 米
设计时间	：2015

Lanzhou

Floor Space	: 10,825 m^2
Gross Floor Area	: 229,700 m^2
Building Height	: 270 m
The Time of Design	: 2015

新古典的高度回归
The Highly Regression of Neoclassicism

“驼铃古道丝绸路，胡马犹闻唐汉风。”

——程祥《“一带一路”交响乐之华彩篇》

在飞去与业主正和交流的途中，这两句诗蓦然心上。建筑师与业主的协作不仅是商业上的合作，在互抒情怀中，思想和文化追求方面都会产生共鸣。在第二次合作中，我们对业主的诉求更加清晰，从建筑表达上溯源文化内涵，思索文脉和业主的需求。

向古典主义致敬的设计尝试

两千年前，金城汤池的丝绸之路，不仅带来茶马互市的传奇，更开启了不同肤色文化的碰撞；2015 年，当兰州需要再竖起一座 270 米的超高层时，我们试图回答什么是这位情怀业主对这座千年古城的期望。项目位于城关区南关十字，正对兰州在建的第一超高层——红楼时代广场，现场已封顶的 266 米高度的压力迎面而来。如果说第一次尝试设计的伦华艺术中心作为现代主义建筑，彰显兰州发展的锐度。这次，我们尝试以回归经典为创作核心，呼唤人们重新理解古城文化的厚度，试图在西方超高层古典风格中找寻文化共通的经典内核。

苛刻环境下的设计挑战

地块现状北侧紧邻军区疗养院是最大的设计难点之一。在这里竖起一座超高层，不仅要顾及日照等规范要求，也要顾及人们的心理感受。设计之初我们用日照软件逆向还原了不影响日照的裙房建筑体量，同时结合体块边界形成退层的强制要求，以整体化的操作方式，针对容量、日照、流线、风格、历史、文化方面多角度“均衡协调”，形成最终的规划方案与形态。

"The camel, the old road and the Silk Road. Hearing the sound of prosperity of Tang and Han Dynasty because of the sound of northern horses."

——Cheng Xiang *Belt and Road*

These two poems rise in the mind on the way to fly to communicate with proprietors. The collaboration between architects and proprietors is not only a commercial cooperation but also a resonance in the ideological and cultural pursuit in the process of expressing their emotions. In the second cooperation, our demands for proprietors were more clear. We traced the cultural connotations from expression of architecture and pondered the needs of the context and proprietors.

The Attempt of Paying Tribute to Classicism

Two thousand years ago, the Silk Road of iron bastions not only brought the legend of the tea-horse trade to each other, but also opened the collision of cultures of different skin colors; When Lanzhou needed to be erected a 270-meter super high-rise building in 2015, we tried to answer what was the expectation that the feeling proprietor had for this millennial city. The project is located in the South Gate Cross of Chengguan District, facing the first super high-rise building that is under construction in Lanzhou-the Red Building Times Square. The on-site roof pressure of 266 meters is on the way. The Lunhua Art Center that is the design of the first attempt, a modernistic architecture, highlight the acuity of Lanzhou. This time, we try to regard the regression of the classic as the core of the creation, calling on people to re-understand the thickness of the culture of the ancient city, trying to find the classic core of the common points of cultures in the Western high-level classical style.

Challenges of Design in the Harsh Environment

One of the biggest difficulties of design is that the north side of the current block is adjacent to the military sanatorium. To erect a super high-rise building here, we must take into account not only the sunshine and other regulatory requirements, but also people's psychological feelings. At the beginning of the design, we used the software of sunshine to reversely restore the podium building mass that does not affect the sunshine and formed the mandatory requirements of the retreat with the boundary of the mass at the same time. After an integrated manner of operation, aiming at "balanced coordination" in terms of capacity, sunshine, streamline, style, history and culture, the final planning scheme and form appears.

主城区核心地段，临政府中轴线
规避苛刻条件、以纯古典主义再演绎经典
融入典雅情怀、以细节诠释黄金比例

The core area of the main city, near the central axis of the government
Avoiding harsh conditions、 reinterpreting classics with pure classicalism
Incorporating elegant feelings、 interpreting golden ratios by details

主体建筑宽高比：1 : 3.8
主体建筑分段比：1 : 2 : 1.2
主体建筑收分比：3 : 1
Aspect Ratio of the Main Building: 1 : 3.8
Segmented Ratio of the Main Building: 1 : 2 : 1.2
Revenue Ratio of the Main Building: 3 : 1

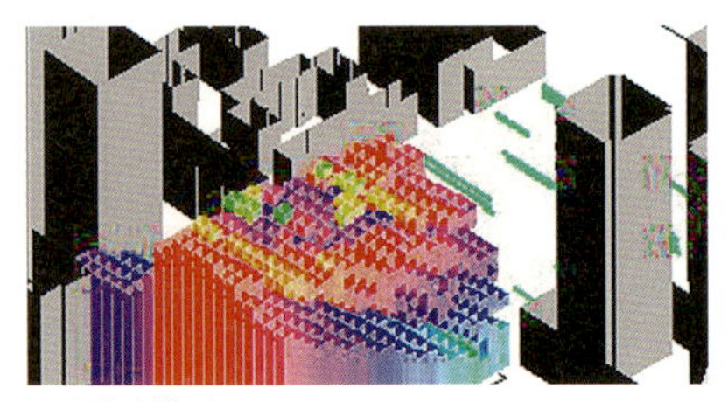

日照逆向模型
Sunshine Inverse Model

36°03′
18.6″ N
103°49′
20.4″ E

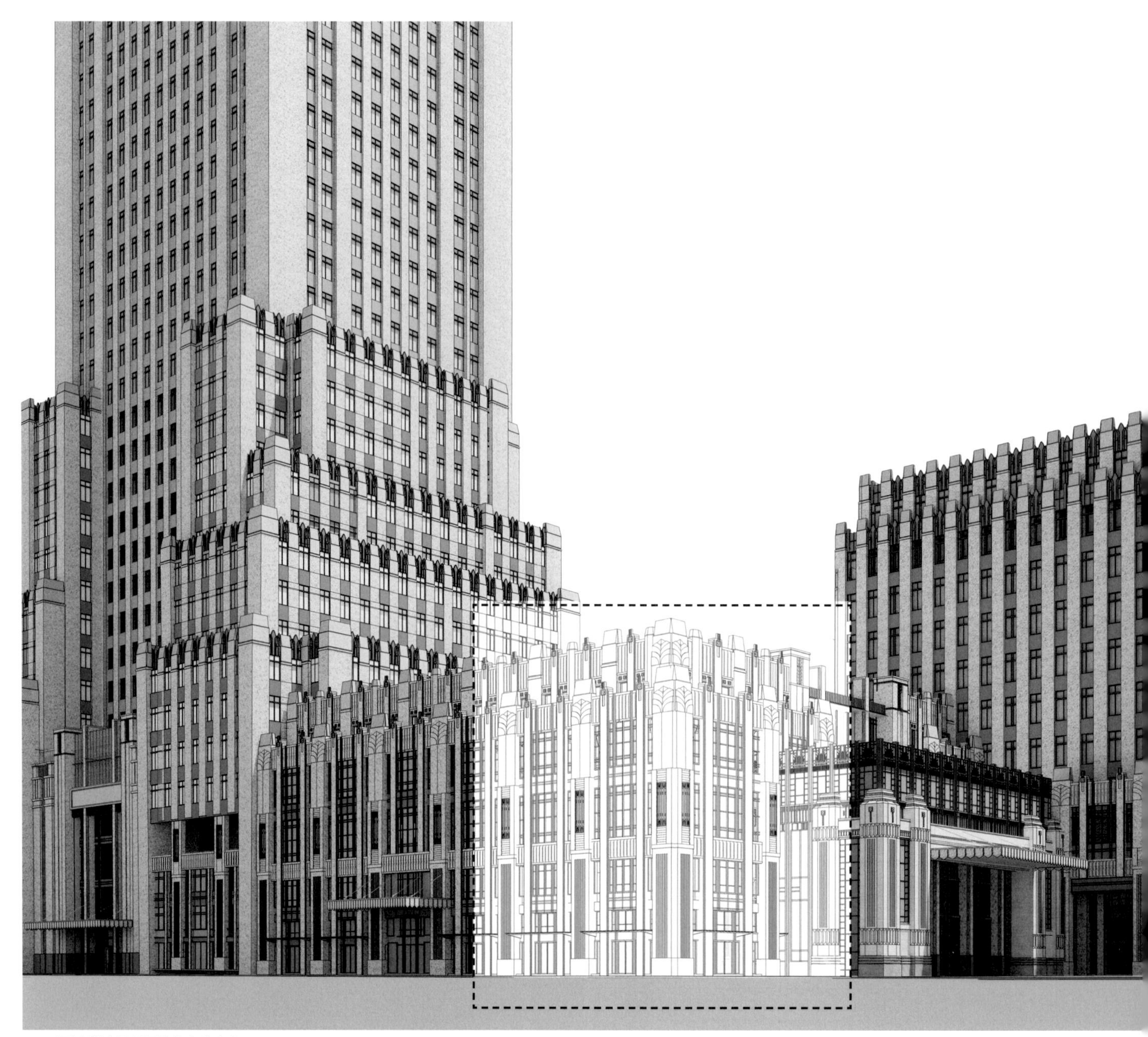

设计尝试呈现极致的古典主义
Design Attempts To Present Extreme Classicism

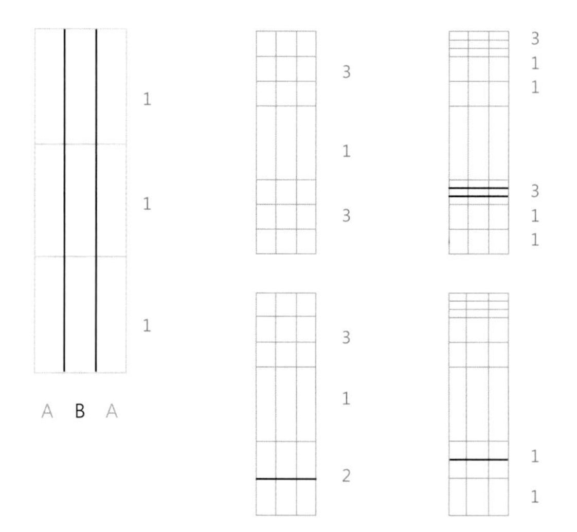

1/3 原则：对称
1/3 Principle: Symmetry

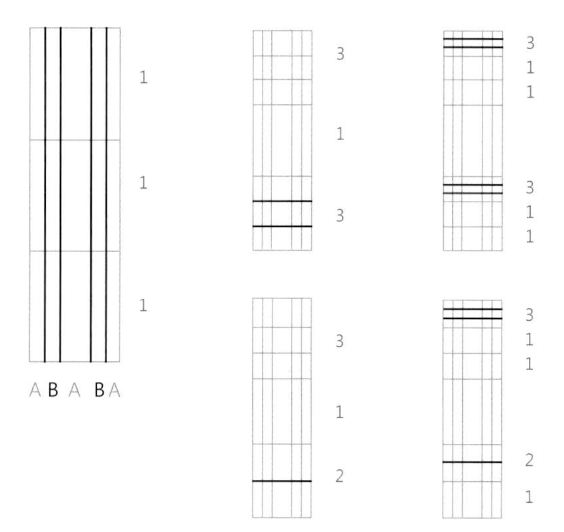

1/2 原则：阵列
1/2 Principle: Array

向经典风格致敬的设计尝试

地块距离伦华中心约 300 米，业主正和集团希望在此再造一个“经典”的新高度，功能设定多家奢侈品品牌进驻。反复探讨后，我们试图将文化与品质交融，尝试在古典中沉淀下经典的内涵，风格选择传统装饰艺术风格，设计以极致的经典美学体现典雅的品质风采。

装饰艺术风格起源于资本扩张后的工业文明，自机械美学孕育，以几何比例控制形体线条，以文化图腾为装饰素材，以比例控制和细节雕饰描绘建筑美。设计之初，我们探究了诸多经典建筑案例与其设计手法，在形体、比例、细节上找寻“**古典设计法**”规则。在设计中，以三模比例控制形体线条，在多维三段式的划分中，以黄金比例推演形体，以 1:3:1 的瘦长比例控制收分，在重复、对称、渐变的法则中推敲立面秩序。此外，设计非常重视对**细节的精致把控**，除了几何美的装饰，细部处理上吸纳了古典符号，加以高度浓缩、简约变形，丰富构件形态，在统一的线条框架中摹画细腻的光影图景。设计希望以这份极致的经典美学，燃起兰州人对这座建筑的自豪。

向经典回归，与城市共鸣出内在的自豪感，正是这个设计真正的价值。即使没有帝国大厦的高度，这份感悟的执着依旧深沉厚重，传承不朽。

能工摹形，巧匠窃意；
切磋琢磨，向道而生。

The Design Attempt of Paying Tribute to the Style of Classic

The plot is about 300 meters away from the center of Lunhua. The proprietor Zhenghe Group hopes to create a new height of "classic" here, setting up a number of luxury brands. After repeated exploration, we tried to blend culture into quality and to settle the classic connotation in the classical style, selecting the traditional Art Deco as the style and embodying the style of elegant quality with the ultimate classic aesthetics.

Art Deco originated from the industrial civilization after the capital expansion, gestating from the mechanical aesthetics, controlling the shape lines with geometrical proportions, using totems of cultures as decorative materials and depicting architectural beauty with proportional control and detailed carving. At the beginning of the design, we explored many classic architectural cases and their techniques of design, looking for the rules of " the method of classical design" in aspects of forms, proportion, and details. In the design, three-dimensional proportions are used to control the lines of the body. In the multi-dimensional and three-segment division, the order of facades is weighed against the rule of repetition, symmetry and gradient by the golden ratio 1:3:1. In addition, the design attaches great importance to the delicate control of details. Except for the decoration of geometrical beauty, detail processing, we have absorbed classical symbols that is highly condensed and simply distorted, enriching patterns of construction drawing delicate light and shadow scenes in a frame of unified line. The design hopes to make people of Lanzhou take pride in this building with the ultimate classic aesthetics.

Returning to the classic and resonating with the city for inner pride is the true value of the design. Even if there is no height of the Empire State Building, the persistence of perceiving that is passing on immortally is still deep and heavy.

Imitative form of skilled workers, crafty craftsmen who are cosy;
Learning from each other and making a living.

解冻历史，再塑品牌 / 上海光明申宏冷库改造

Thawing History, Remolding Brands / The Reconstruction of Bright Shenhong Cold Storage in Shanghai

上海

占地面积 ： 27 400 平方米

建筑面积 ： 75 072 平方米

建筑高度 ： 28 米

设计时间 ： 2016

Shanghai

Floor Space : 27,400 m^2

Gross Floor Area : 75,072 m^2

Building Height : 28 m

The Time of Design : 2016

远东第一冷库蜕变
Far East First Cold Storage Transmutation

历史在被文明挖掘出时最为耀眼，设计正是为这份辉煌提供看台。三栋列为市级文保建筑的冷库，一家横跨半个多世纪的民族品牌企业，中环高架下的一处不大的地块，一场与国际大师合作的概念设计就此展开。我们面对的不仅是一次建筑概念方案设计，也是一轮与历史文化、城市建筑、产业品牌的对话。

History is most dazzling when it is excavated by civilization. Design is to provide a stand for the glory.There are three cold storages listed as the city-level conservation buildings, a enterprise of the national brand that spans more than half a century, a small plot underneath the Zhonghuan elevated bridge and a conceptual design that cooperates with international masters. We are faced with not only a project design of architectural concepts, but also a dialogue with historical culture, urban architectures and industrial brands.

解冻沉浸于温度的历史

Thawing the History that is Immersed in Temperature

50 年前，这里还是荒芜一片。1966 年，容量 1 万吨的老库建成投产，两年以后，容量 8 000 吨的新库在其北侧拔地而起，占地 55 亩、冷冻品总储存量将近 2 万吨的冷库，是当时远东地区最大的冷库。这里曾维系全上海冻品 65% 的供应量，这里曾见证光明这个民族食品品牌走向现代化、综合化、国际化的历程。半个多世纪过去了，80 吨液氨已全部运出，申宏冷库的历史使命已安然卸下，地块渐渐归于沉默，仅留下三个封闭的混凝土巨构，等待在新时代涅槃。

The place was still ridiculous 50 years ago. In 1966, an old warehouse with a capacity of 1,000 tons was completed and put into operation. Two years later, a new warehouse with a capacity of 8,000 tons was erected on the north side of the old warehouse, covering a total area of 55 acres. The cold storage with a total capacity of nearly 20,000 tons was the largest in the Far East. 65% of the total supply of frozen products in Shanghai was maintained here. The place have witnessed the development of modernization, totalization and internationalization of the national food brand of Bright. More than half a century has passed, 80 tons of liquid oxygen have all been shipped out. The historical mission of Shenhong Cold Storage has been safely completed, the land gradually becoming silent, only three closed macrostructure of concrete left, waiting for a new era of nirvana.

在这个消费观念日新月异的时代，伴随消费方式升级，食品产业与物流、商业复合，城市服务业态更新，新的商业机遇正以前所未有的各种方式发生。2018 年，上海将首次举办国际高端食品饮料与进出口食品展览会，而光明也以申宏冷库等项目的改造为契机，融入这一轮商业机遇，让品牌在更广泛的认知中凝聚更强的价值，让服务在经营与口碑的双丰收中驱动产业进一步升级。

In the era of ever-changing consumption concept, with the upgrading of patterns of consumption, the food industry, logistics and business combine and types of operation of the urban service are updated. New opportunities of business are taking place in unprecedented ways. In 2018, the first international exhibition of high-end food and beverage and imported food will be hold in Shanghai. Bright will also regard the transformation of Shenhong Cold Storage and other projects as an chance to integrate into this round of business opportunities so that the brand will gather more strong values in a broader sense and industries will be further upgraded by services in the double harvest of operation and public praise.

历史随温度凝结，沉淀在文化与记忆中，成为冷库这个符号下的内涵。而设计则以阐释这个最鲜明的符号——冷库——为起点，逐一展开。

History condenses with temperature, settling in cultures and memories and becoming the connotation of the symbol of cold storage. The design is based on the interpretation of the most distinctive symbol, cold storage, and starts one by one.

上海中环、邻上理工
五十余年历史、市文保建筑、结构特殊、体量巨大
产业升级、肩负来自世界级食品集团的发展雄心

The middle ring road of Shanghai, adjacent to University of Shanghai for Science and Technology
More than 50-year history, the city-level conservation architecture, special structures and huge volume
Industrial upgrading, development ambition that comes from the world-class food group

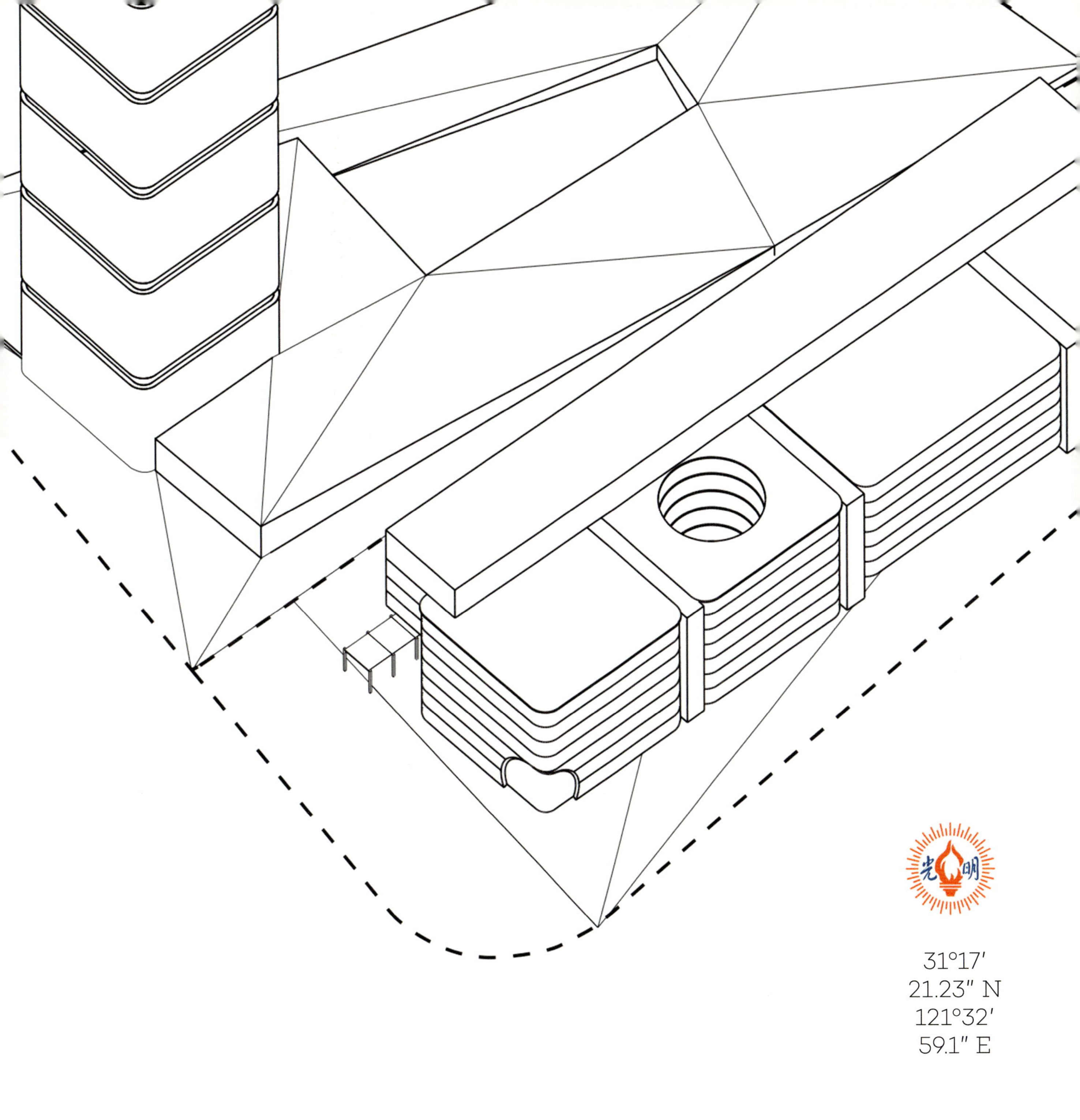
光明
31°17′
21.23″ N
121°32′
59.1″ E

如何承载光明转型发展的示范性？
How to Bear the Weight of the Demonstration of the Transformation and Development of Bright?

申宏冷库改造——保护“冰箱”计划 - 光明垂直森邻

Customized Reform Design for Shenghong Cold Storage

前期概念设计合作方：
Partners of Pre-design of Concept

方案合作：博埃里工作室，意大利
Program Cooperation: Boeri Studio, Italy

策划顾问：颢兴茂文化创意发展有限公司，上海
Planning Consultant: Haoxingmao Limited Company, Shanghai

品牌顾问：肯默整合设计，台北
Brand Consultant: CANMORE Design, Taibei

视频制作：骑鲸客文化，上海
Video Production: Qijiki, Shanghai

文保研究：明悦设计事务所，上海
Study of Conservation: Mingyue Design, Shanghai

规划顾问：上海市规划设计研究院
Programme Consultant: Shanghai Institute of Programme and Design

视觉顾问：今尚视觉，上海
Optesthesia Consultant: Gingshang Optesthesia, Shanghai

地块分析
Block Analysis

冷库 Cold Storage　低层 Low　场地 Site

外科手术式的改造

该地块形状较为特殊：一期用地狭长，长 330 米，最宽处仅 80 米，长宽比不足 1:4。不仅如此，地块被一条 20 米宽规划道路划分为南北两块，南地块被冷库切分为三段；北地块较规整，但可新建面积仅占总用地 1/5。面对军工路，设计不仅需考量视线被高架遮挡后的形象问题，依据上位规划要求有 20 米宽的绿化退界，建设范围被进一步压缩。

地块内原先有大量临时市场及设备用房，拥簇着冷库主体建筑。三座冷库是三个高约 30 米、长宽约 37 米的混凝土建筑，外墙厚 1.5 米，开窗面积不足 0.1%，由于文保条例要求，外墙立面不得改动，这就导致如果不通过人工照明，内部将是完全封闭的黑空间，这与偏公共性的功能设定存在有矛盾。冷库内部是 6 米柱网下的无梁楼板结构，层高为 4~4.5 米不等，留有设备、管道的遗构，让空间更加幽暗而狭促。从发黄的设计存档图纸中可以看出，这些冷库是特殊时期为了储存物资而兴建，并不适合新时代背景下功能的使用要求。

Surgical Reconstruction

The shape of the plot is rather special: The land of the first phase is long and narrow, with a length of 330 meters and 80 meters at the widest point, whose aspect ratio is less than 1:4. Nay, the plot was divided into two blocks, the north and the south, by a 20-meter-wide planning road. The south block was divided into three sections by the cold storage. The north block was more regular, but the newly-built area accounted for only one-fifth of the total land. Facing the military road, the design needs to consider the image problem that the line of sight is obstructed by elevated bridge. According to the superior planning requirements of a setback line of greening of 20-meter width, the scope of construction is further compressed.

There were a large number of temporary markets and equipment rooms in the plot and the main building of the cold storage was clustered. The three cold storages are three concrete buildings with a height of about 30 meters and a length and breadth of about 37 meters. The thickness of outer wall is 1.5 meters and the area of opened window is less than 0.1%. Due to the requirements of conservation regulations, the facades of the external walls must not be altered, which leads that the interior will be a completely closed black space without artificial lighting, which is in contradiction with the setting of partially public functions. The interior of the cold storage is a beamless floor structure under a 6-meter column grid, with a floor height of 4m to 4.5m. The left structure of equipment and pipes make the space more dim and narrow. It can be seen from the archived and yellowed drawing of the design that these cold storages that were built for the purpose of storing materials in a special period are not suitable for the use requirements of the functions in the new era.

场地现状
The Situation of the Site

在内外空间都极为局促的环境下，设计首要解决的就是空间发展的策略问题。由于规划条件的约束，让空间向上发展几乎是唯一的选择。我们通过前后七轮对场地的考察，最后确立了“外科手术式”的改造策略，以最大可能保存建筑的整体性，将新体量从顶部置入，结合屋顶的生态农场及餐饮业态置入，形成“空中花园餐厅”，同时顶部“光廊”结合以 LED 技术形成夜间中环的壮丽标识。

以体量对话城市，形成整体性与地标性的印象。以空间叙事内容，创造独有的体验与场所印记。以文化链接企业，塑造强烈的品牌价值感与企业价值观。我们尝试在产业导入的基础上结合以上三点，完成新旧建筑的统一与场地的再生。

In an environment where both interior and exterior spaces are extremely cramped, the primary solution to the problem of design is the strategy of spatial development. Due to the constraint of conditions of program, it is nearly the only option to make space develop upwards. After examining the site seven rounds, we finally established a "surgical" strategy of construction to preserve the integrity of the building with the greatest possibility, putting new volume from the top, combining the ecological farm and restaurant status of the roof, which forms "aerial garden restaurant". At the same time, the top "light gallery" that is combined with LED technology forms a magnificent logo of middle ring at night.

Talking with cities by volume is to form an overall and landmark impression. A unique experience and place stamp is created with narrative content of space. Linking companies with cultures is to create a strong sense of brand value and corporate value. We have tried to combine the above three points on the basis of industry introduction to complete the unification of the old and new buildings and the regeneration of the site.

Step1

建筑面积 38 095 m²

容积率 1.39

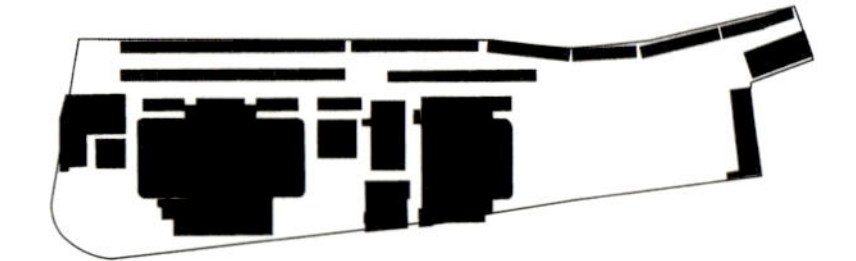

Step2

拆除低层和多层建筑

冷库的交通体

Dismantling Buildings

保留面积 25 200 m²

容积率 0.92

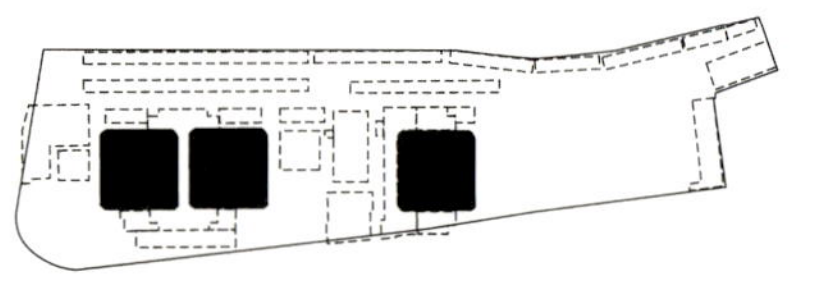

Step3

增加一栋玻璃体

一栋公寓和商业街

Addition Buildings

建筑面积 75 200 m²

容积率 2.74

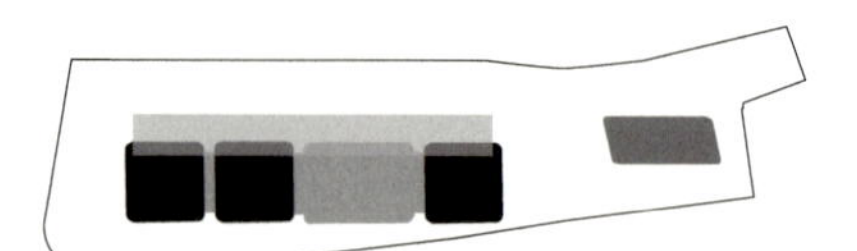

本身的混凝土体量如同巨山便是其价值，去接受并强化而不是试图去破坏或毁掉的策略。

The volume of concrete itself is like a giant mountain. It is its value to accept and strengthen rather than try to destroy or destroy the strategy.

Original
Condition&
Staging range

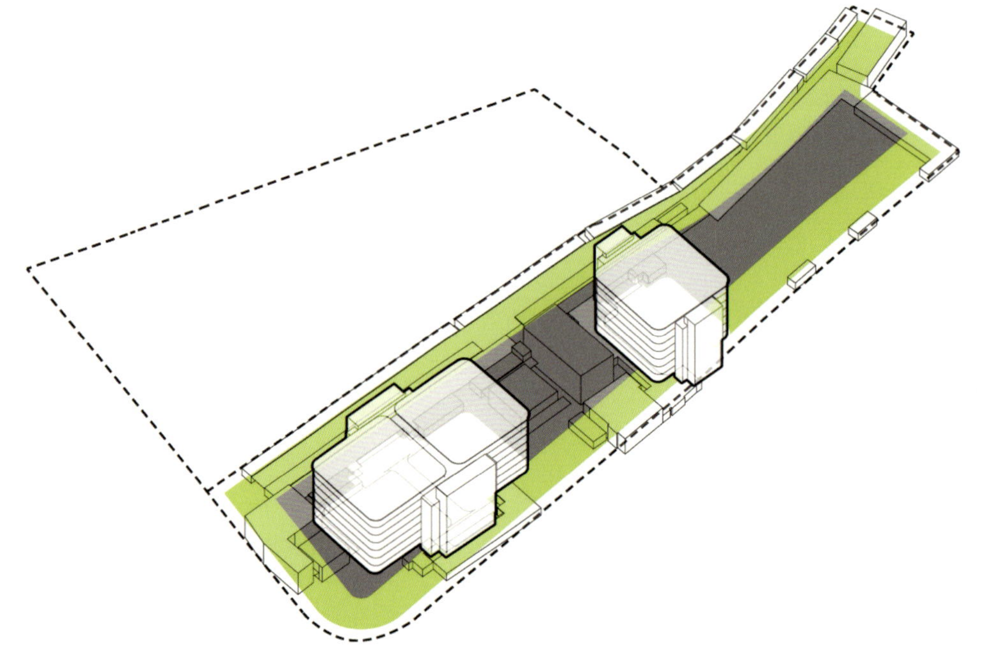

Total 总体　FAR 3.37

地上总建筑面积：160 800 ㎡

地下总建筑面积：112 500 ㎡

建筑密度：32%

Total Floor Area On The Ground: 160,800 ㎡

Total Floor Area Of Underground: 112,500 ㎡

Building Coverage Ratio: 32%

Phase I 一期　FAR 2.74

地上总建筑面积：75 200 ㎡

地下总建筑面积：52 500 ㎡

建筑密度：32%

Total Floor Area On The Ground: 75,200 ㎡

Total Floor Area Of Underground: 52,500 ㎡

Building Coverage Ratio: 32%

Phase II 二期　FAR 4.22

地上总建筑面积：85 600 ㎡

地下总建筑面积：60 000 ㎡

建筑密度：32%

Total Floor Area On The Ground: 85,600 ㎡

Total Floor Area Of Underground: 60,000 ㎡

Building Coverage Ratio: 32%

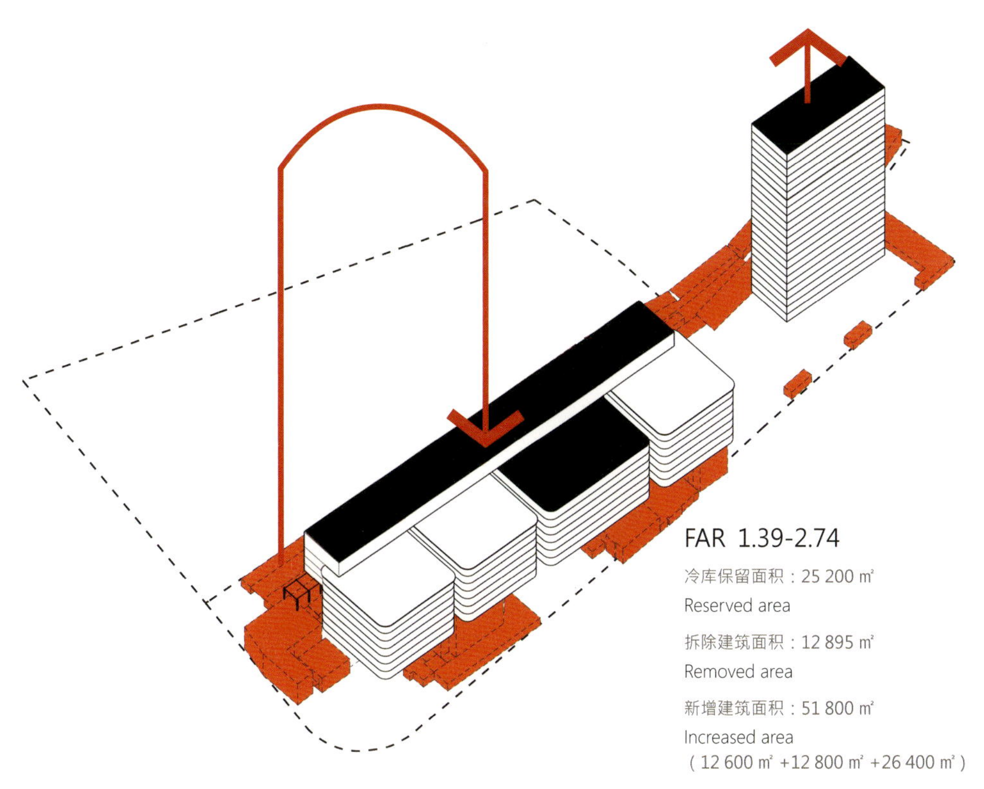

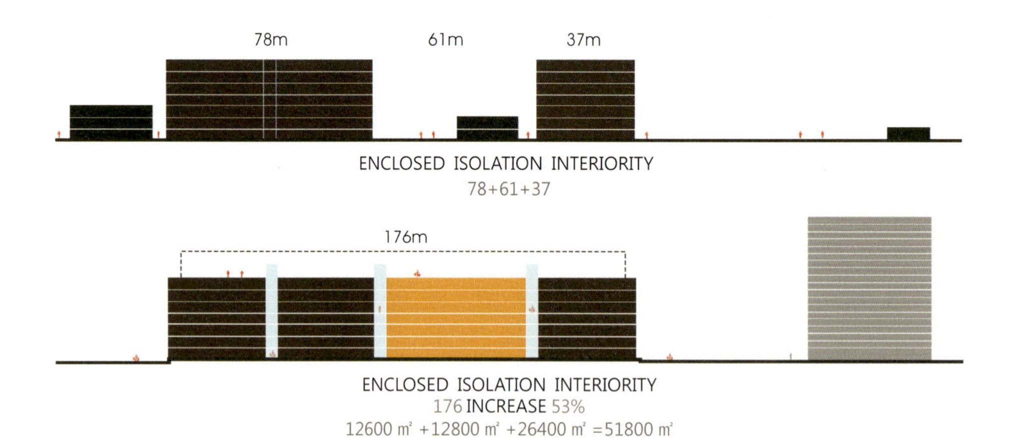
78m
61m
37m
ENCLOSED ISOLATION INTERIORITY
78+61+37
176m
ENCLOSED ISOLATION INTERIORITY
176 INCREASE 53%
12600 m² +12800 m² +26400 m² =51800 m²

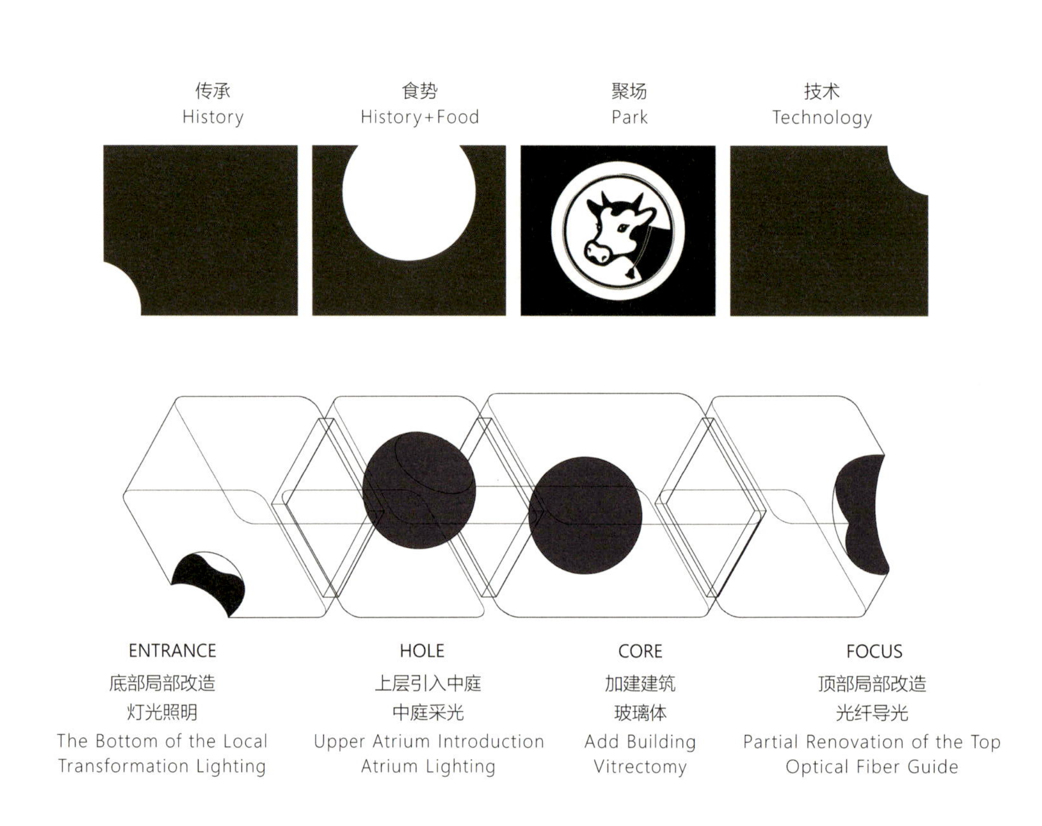
传承
History
食势
History+Food
聚场
Park
技术
Technology
ENTRANCE
底部局部改造
灯光照明
The Bottom of the Local
Transformation Lighting
HOLE
上层引入中庭
中庭采光
Upper Atrium Introduction
Atrium Lighting
CORE
加建建筑
玻璃体
Add Building
Vitrectomy
FOCUS
顶部局部改造
光纤导光
Partial Renovation of the Top
Optical Fiber Guide

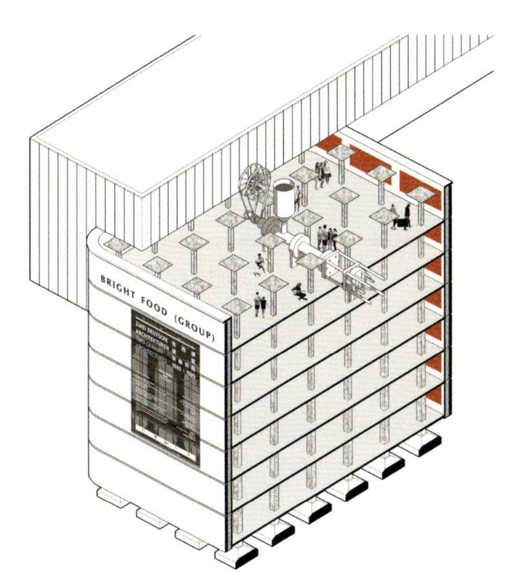

工业遗存将成为空间
Industrial Legacy Will Become the Space

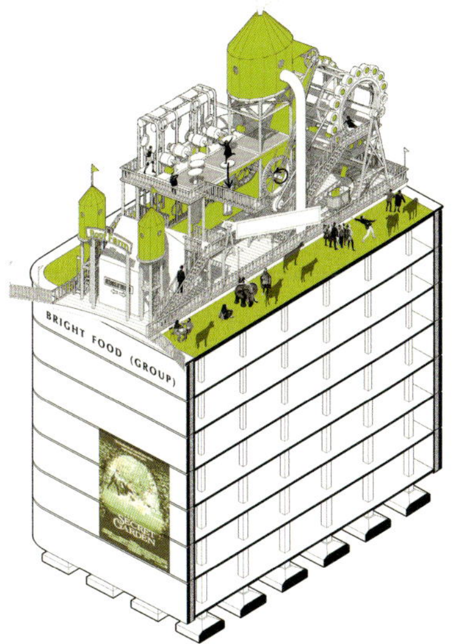

立体、多元的城市生态农场
Tridimensional and Diversified Ecological Farm of Cities

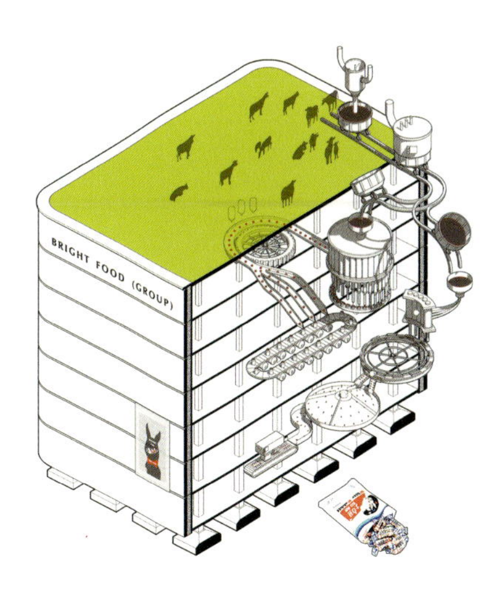

见证一颗奶糖的诞生
Witnessing the Birth of a Toffee

第四座“冷库”、四个跳跃的“起司球”与一条天空光廊

在一期设计中，场地的联系与冷库的呈现形式尤其难处理。在确立了尊重冷库、新旧对话、激活场地三大设计原则后，设计尝试随即展开。

在拆除配建的临时用房后，设计了第四座“冷库”，一个与冷库等高的玻璃虚体，让场地在沿军工路高架界面上获得更强的整体性。同时，也让新功能与体验空间在容量上得到平衡。玻璃体底层架空，在两栋冷库间形成广场，创造内外空间交互的节点。如同冰砖的外立面与冷库的混凝土材质形成对比，在环境的渲染下熠熠生辉，形成强对比话题点的同时，商业功能的适用性也随之提升。

此外，为了创造更丰富的空间交流，设计在方形体块中插入了四个球体，并与外立面相互融合，形成独有的标志。结合高架上的城市视角，将最北端冷库角部挖去一个 1/4 球体的容量，形成朝向天空的光点。结合街角的行人视角，让最南部冷库在角部打开一个 1/4 球体的立面，形成整个序列的入口空间。置换中间的混凝土冷库结构，并塑造一个半球状中庭，引入光线。在玻璃“冷库”的体量中悬浮一个封闭的球体体量，内置新颖的体验功能，塑造形象上的视觉焦点。四个“起司球”与方正的形体形成戏剧化的对比，让严肃冷峻的工业遗构有了轻松的形象话题。

基于整体性考虑，我们设计将“天空光廊”在背部与冷库相互咬合。光廊中冷库的外墙将被掀起，断裂中暴露出 1.5 米墙厚与盘布缠绕的设备管道，让城市界面的路人能够感知这里曾被温度标记的印记。

The Fourth "Cold Storage", Four jumping "guitar balls" and a Celestial Spotlight Gallery

In the first phase, connection of sites and form of presentation of the storage are particularly difficult to handle. After establishing three major design principles of respecting the storage, old and new dialogues and activation of sites, the attempt of design begins.

After dismantling equipped and temporary housing, we designed the fourth "storage". An invisible glass body with the same height as the storage makes the site more integrated in the elevated interface along military road. Meanwhile, new functions and experience space are also balanced in terms of capacity. The vitreous body is overhead, forming a square between two storages to create nodes for the interaction of internal and external spaces. The facade of ice brick is contrasted with concrete material of the storage, shining under rendering of environment and forming a topic that is strongly contrasting. Meanwhile, applicability of functions of business also increases.

Besides, to create a richer space for communication, the design incorporates four spheres in a square block and merges with the facade, which forms a unique logo. Combined with city perspective in elevated bridges, the area of the corner of the northernmost storage will be dug out of 1/4 sphere to form a light point that faces the sky. According to pedestrian perspective of the street corner, the southernmost storage is opened a facade of 1/4 sphere at the corner to form the space of entrance of entire sequence. The structure of intermediate concrete storage is replaced and a hemispherical atrium is created to introduce ray of light. A closed volume of spheres is suspended in the volume of the vitreous "storage", with novel functions of experience that are built-in to shape the visual focus on the image. The dramatic contrast between the four "Cheese balls" and the form of founder has given the solemn and cold industrial relics a relaxing topic.
Based on holistic considerations, we make the back of "sky light gallery" mesh with the storage. External walls in the gallery will be lifted. The wall whose thickness is 1.5m and dish-wrapped pipes of equipment are exposed in the break. The passers-by at the city interface can perceive the mark that was once marked by temperature.

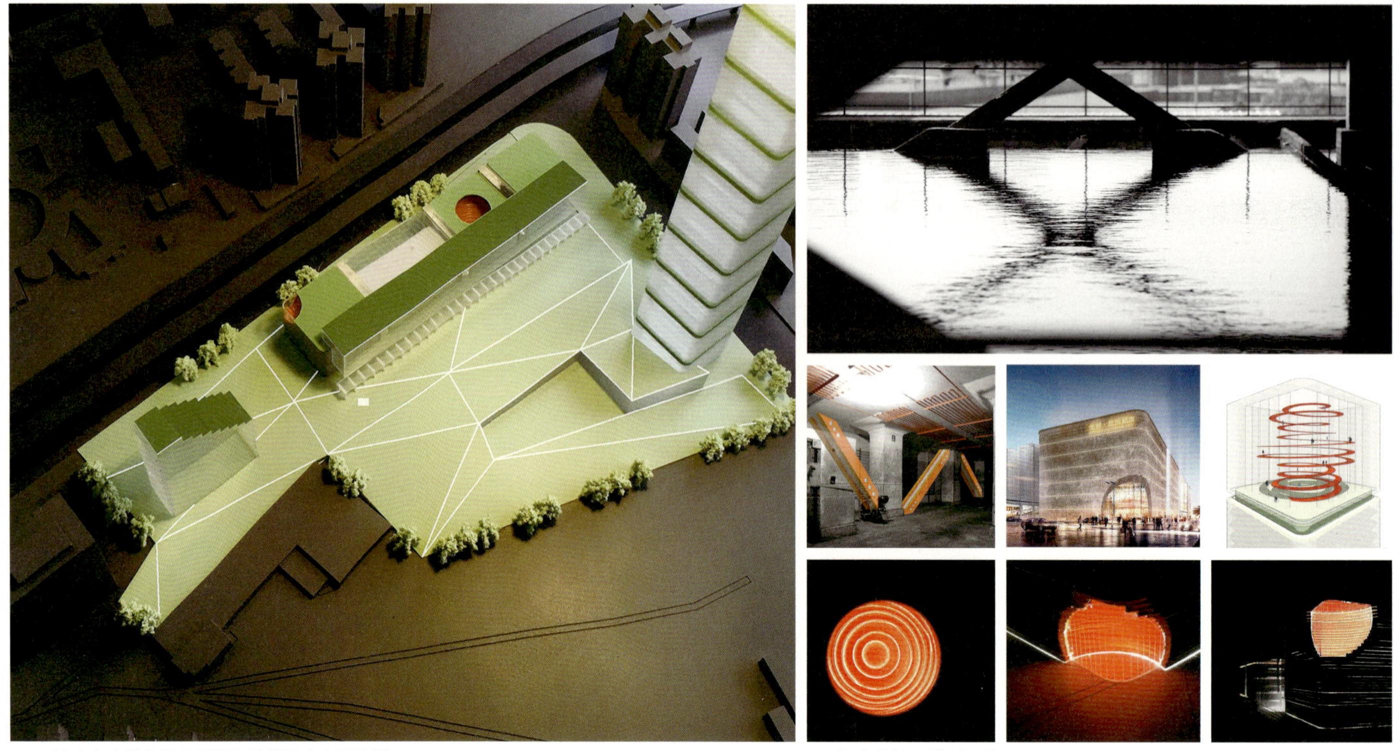

结合农产品交易大厅设计的超尺度农场公园
A Super Scale Farm Park Designed With Agricultural Products Trading Floor

概念分析 + 模型照片
Concept Analysis + Model Photos

新高度再塑品牌

冷库不单是敦实的巨体，也是一个时代精神的包容。设计不仅要解决空间的发展，更要传播并重塑光明品牌的认知。光明为人们所熟识的是乳业，但它远不止于此。自益民食品一厂发迹至今，它包囊了农林牧副渔等传统产业与城市物流、地产开发等综合产业，它涵盖了海博、农工商、冠生园、正广和、梅林、光明乳业等一系列知名品牌。它既如一位长者，经历无数风雨，又如一位学子，在新时代整装待发。

我们在参观相关光明的所有博物馆，零距离感受光明的历史后，逐渐形成以新业态探索产业联合、以新高度标识品牌力的想法。一期作为展示媒体，植入了光明品牌陈列、体验、市集、研发等内容，充分展示光明产业的广度，并在与城市互动中探索产业研发、消费体验与社会效能的聚核形态。

Remaking Brand to a New Height

The cold storage is not only a real giant, but also a pardon of the spirit of eras. Design not only address the development of space, but also propagate and reshape the perception of the brand, Bright. The dairy industry of Bright is well-know to people. But it is far more than that. Since Yimin's No.1 plant of food was founded, it has covered traditional industries such as agriculture, forestry, animal husbandry, fishery, as well as urban logistics, the development of real estate and other comprehensive industries, which covers Haibo, Nonggongshang, Guanshengyuan, Zhengguanghe, Meilin, Bright Emulsion and other series of well-know brands. It is like an elderly person, experiencing countless wind and rain, and a student who is ready to go in the new era.

After visiting all the museums of Bright and experiencing the history of Bright from zero distances, we have gradually formed the idea of exploring industrial alliances by new formats and identifying power of the brand at a new height. The first phase, as a media of displaying, was embedded with display, experiences, markets, research and development of the brand Bright, fully demonstrating the breadth of the Bright industry. In the process of interacting with cities, the polynuclear form of industrial R&D, consumer experience and social effectiveness was explored.

冷库内部（关停后）
Inside the Refrigerator (After Closing)

城市天际中亮红一抹：光明火炬
Bright Red In the City Skyline: The Torch of Light

沿中环线的壮丽街景——巨幅 LED
The Magnificent Scene of the Street Along the Line of Middle Ring - Giant LED

二期的地景式设计——最大尺度的开放性
The Second Phase of the Design in the Type of Landscape - the Largest Scale of Openness

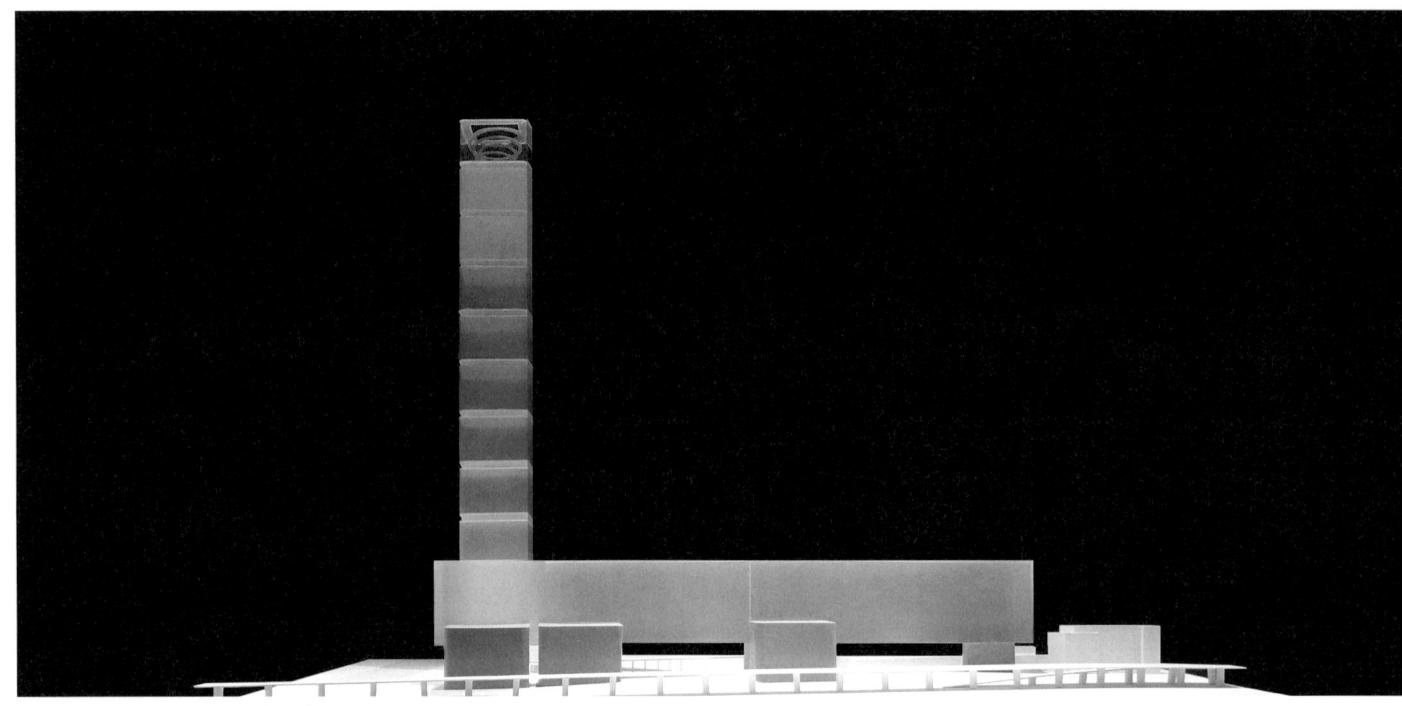

早期工作模型：不融化的“冰砖”
Early Working Model: Thawless "Ice-cream Brick"

一座城市的高度，一个品牌的广度。
The Height of a City , the Extent of a Brand.

冷却塔建造于 1940 年，为木质外层，水泥底座，结构独特，是当时益民食品一厂的标志性建筑之一。
The cooling tower, built in 1940, whose outer layer is wooden, cement base is concrete and structure is unique, was one of the landmark buildings of the Yimin Food No.1 Factory.

后期工作模型：光芒飘舞
Later Work Model: Light Wave

二期作为光明在上海的地标，以一个超尺度的农场公园与一座超高层综合体标识光明形象，塑造理想的城市模型。270 米超高层的九节塔造型源于益民食品一厂博物馆的镇馆之宝——建国之初的木构冷却塔，每节轮廓与一期冷库一致，并在顶部通过观光坡道构筑出火炬的形象，以城市天际中亮红一抹，象征着品牌精神。在低层部分，设计将农产品交易大厅等功能以地景建筑的形式呈现，将地表开发密度最大程度的降低，形成 20 000 平方米的超尺度生态农场公园，占用地面积近 70%，以生态发展的价值观改善城市品质。

可以说，一期是产业聚核的展示，二期是品牌在城市中的标杆，设计以新高度彰显光明品牌的新姿态。

The second period, as a surface of Bright in Shanghai, mark the image of Bright with a super-scale farm park and a super high-rise complex, shaping the project's urban model. The modelling of 270m high-rise nine-section tower originates from the treasure of the museum of the Yimin Food No. 1 Factory - the wooden cooling tower at the beginning of our nation, whose contour of each section is same as the first-stage cold storage. The image of torch is constructed on the top through a sightseeing ramp, a red touch in the city skyline, symbolizing the spirit of the brand. At the lower level, the design of agricultural trading halls and other functions is presented in the form of landscaping buildings, which reduces the density of surface development to a minimum, forming a super-scale ecological farm park that is 20,000 square meters, occupying an area of nearly 70%, with the value of ecological development to improve the quality of the city.

It can be said that the first phase is the display of industrial polynuclearization and the second phase is the benchmark of the brand in the city. The design highlights the new posture of the brand Bright with a new height.

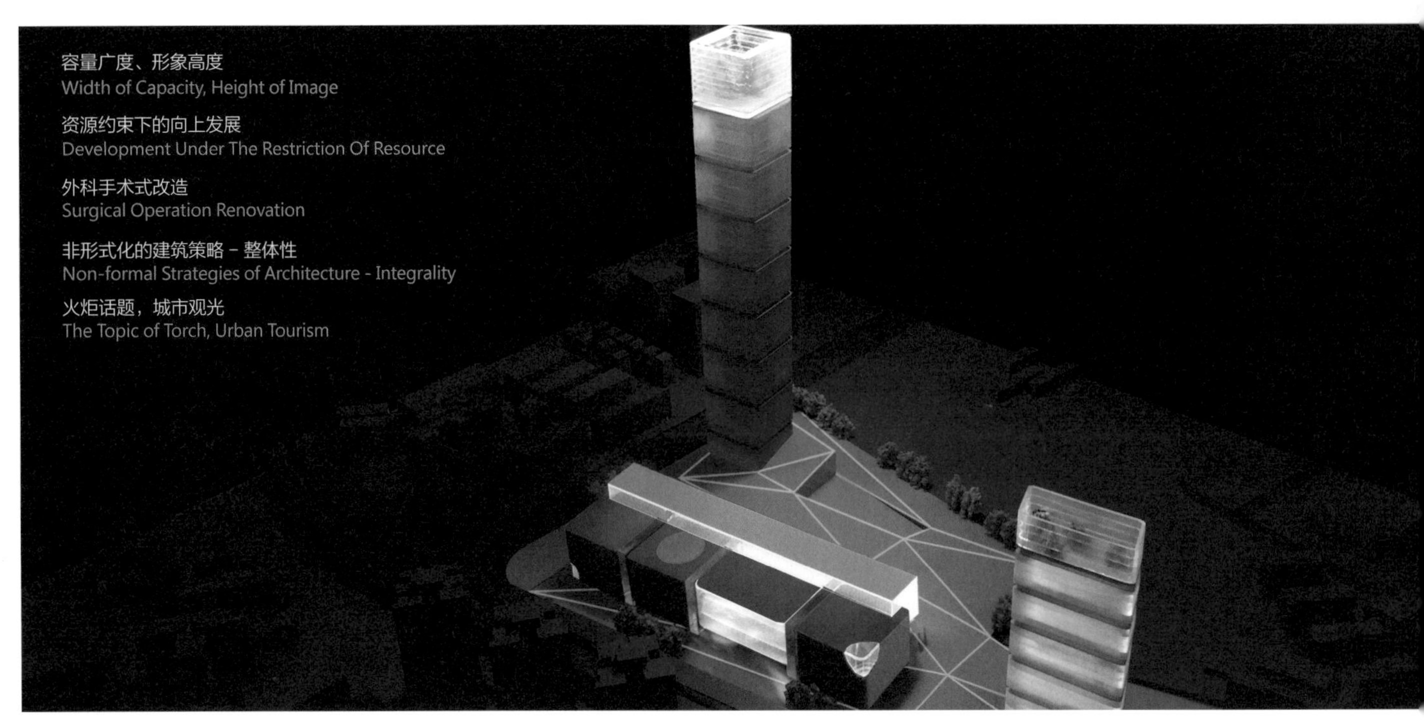

点亮光明，于历史记忆中重现新高度与新亮度
Lighting up and Recreating New Height and Brightness in the Memory of History

对团队来说，还有一段特别的项目故事，在最终汇报前一晚凌晨，工作完成后我们再次造访了地块，我们想再来感受一下这几栋建筑。几个月的工作与调研让我们对这个项目产生了浓厚的感情。想办法叩开南侧住宅门禁后，我们爬上了楼顶，第一次感受地块的深夜，中环上的车寥寥无几，地块内几乎没有一丝灯光，想象我们的方案塑造的光明……寒风将双颊吹得干裂，但内心却无比的炽热。第二天早上，我们将对"光明"与冷库地块的感动融入汇报，让场内听众都有不小的触动。业主不仅认可了我们方案，更认可了我们团队，在之后赢得多次设计任务委托。

设计之外的故事，也是故事之内的设计。

For the team, there is a special story of project. In the early morning before the final report, we visited the plot again after the completion of the work. We would like to feel the buildings again. A few months of work and research gave us a strong feeling for the project. After trying to open the door of the southern house, we climbed to the top of the building, feeling the night of the plot for the first time. There were few cars on the middle ring and there was almost no light in the plot. Let us Imagine the bright light that is made by our plan. The cold wind blows the cheeks dry, but the heart is extremely hot. The next morning, we will integrate the affection of "Bright" and the plot of cold storage into the report. Every listener on the spot will be touched. Bright recognized not only our plan, but also our team. We won many assignments of design later.

The story outside the design is also the design within the story.

博埃里与垂直森林：植物学家选取野外灌木
Boeri and Vertical Forest: Selecting Field Shrubs by Botanists

2017 年 2 月 寒冷深夜中与冷库的对话
Conversations With Cold Storage,2017,2

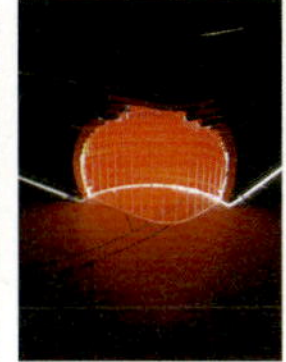

2016 年 11 月 汇报前一日，水石内部讨论
The Day Before Reporting,2016,11

2017 年 2 月 博埃里汇报垂直森林方案
Report of the Plan of Vertical Forest by Boeri,2017,2

与国际大师合作和博弈

这次设计是对团队的一次挑战——与国际大师的合作和博弈。斯特凡诺 · 博埃里，作为米兰垂直森林的缔造者、*Domus* 杂志主编、米兰世博会总建筑师，在行业内有不小的影响力。在设计进程中，我们非常敬佩对方独到的设计与眼光，在城市、生态与建筑的结合中，博埃里提出的“城市生物多样化”具备很强的前瞻性。但是，由于城市的差异，以垂直森林落位上海来看，高土地价值、季节性台风、高成本维护 [1] 均是不利要素。

这是一个历史建筑的改造项目。我们认为，旧建筑特质、品牌精神、产业定位是重要的关注要素。所以，设计选择将冷库作为地块的焦点，有策略地对旧建筑保护、还原与改造，在“有所不为”原则上适度强化设计的锐度。新建筑遵循了同样的原则，通过形体、“火炬” [2] 元素、多重“可食性” [3] 等方面研究，以简约而不简单的形式既衬显了旧建筑，又塑造了“新高度”。正是因为历史、品牌、生态、产业四方面的融入，我们的设计最终呈现出“高度”的定制性，获得了业主的高度认可。

Cooperating and Gaming With International Masters

This design is a challenge for the team-cooperation and game with International masters. Stefano Boeri, as the founder of Milan's vertical forest, editor-in-chief of Domus, and chief architect of the Milan Expo, has a lot of influence in the industry. In the process of designing, we greatly admire the unique design and perspectives of the other party. In the combination of cities, ecology, and architecture, Boeri's "ruban biodiversity" is highly forward-looking. However, due to the differences from cities, considering vertical landfall in Shanghai, high value of land, seasonal typhoon and high cost of maintenance are all unfavorable factors.

This is a reconstruction project of historic buildings. We believe that trait pf the old buildings, the spirit of the brand and industry positioning are important elements. Therefore, the design choose the cold storage as the Focal point of the plot. It will strategically protect, restore and transform the old building and appropriately strengthen the design's sharpness on the principle of doing nothing. The new building follows the same principle. It was studied by the form, the torch element, multiple edibleness and other aspects. It highlights the old building and shape the new height in a simple but not easy form. It is precisely because of the integration of history, brand, ecology, and industry that our designs ultimately show a high degree of customization and have been highly recognized by proprietors.

设计过程模型：梦舞台与天空城
The Model of the Process of Designing: Dream Stage and Sky City

设计过程模型：冰峰
The Model of the Process of Designing: Ice Peak

手工模型
Manual Model

龚兵 马如柏
Bing Gong / Rubai Ma

玻璃、木、瓦楞纸板、三氯甲烷胶
Glass, Wood, Corrugated Board
Trichloromethane Gum
1 : 750

过去的文本，现今的“模型”
The Text of the Past, The Modern "Model"

新旧建筑关系
Relationship between New and Old Buildings

模型推敲过程：折纸研究天际线

The Process of the Scrutiny of Model: Study the Skyline by Paper Folding

手工模型

Manual Model

于洪浩 段嫣然

Honghao Yu / Yanran Duan

涂布纸

Coated paper

1 : 100

东方"船说"的故事 / 东方盛泰澜金泰湾度假酒店[1]（五星）

The Story of "Ship" in the East / Dongfang Shengtailan Jintai Bay Resort Hotel (Five Star)

东方

占地面积	：37 089 平方米
建筑面积	：116 097 平方米
容积率	：2.5
设计时间	：2016

Dongfang

Floor Space	: 37,089 m^2
Gross Floor Area	: 116,097 m^2
Plot Ratio	: 2.5
The Time of Design	: 2016

海南首座原生亲子主题

The First Native Parent-child Theme in Hainan

大海是憧憬，是一条柔软的地平线，是一面巨大的镜子。面对海，我们内心最柔软与最坚硬的部分随着海浪上下起伏，就像孩子参加故事会的心情．怀揣着对海的纯真遐想，我们在海南省东方市设计了一座“特别特色”的酒店。

东方市位于海南西海岸，是海南第三大滨海城市，港口众多、对越贸易历史久远。东方的名气虽不比三亚、海口，但有独特的原住民文化与原始生态资源，物华天宝，奇珍异禽、沉香花梨盛产于此。相比周边城市普遍过度商业化发展，东方原生态的本土资源具备很大的旅游发展潜力。

项目位于东方市板桥镇的海边，周边原生态滩涂环绕，往东远观尖峰山，往西海景视野开阔。作为一处高品质的五星级度假酒店定位，地块周边现状发展仍不够成熟，宏观上：最近的机场只能借道三亚或海口，需一个多小时车程；周边道路以镇级公路为主，窄而旧；附件景区开发水平较低，周边配套品质不高。微观上：用地规模较小，形状呈不规则扇形，规划中住宅环绕；地块距海 300 米，潟湖、防波堤阻隔的原因，步行不可达；项目容积率 3.5，除 400 间客房外，另有约 1000 间海景公寓的设计要求。在以上诸难点下，想获得设计和经营上的成功，对我们和管理公司均提出了不低的要求。

传统的设计思路显然不行。我们不仅需要在文化、经济、景观等资源深度挖掘，在效率方面满足管理经营要求，更重要的是在独特吸引力个性方面的塑造——而这，可能正是本次设计最大的难点。

The sea is longing, a soft horizon and a huge mirror. Facing the sea, the softest and hardest part of our hearts goes up and down with the waves, just like the children’s mood of participating in the story meeting. With Innocent delusions, we designed a "specially characteristic" hotel in the Dongfang, Hainan.

Located on the west coast of Hainan, Dongfang where there are many ports is the third largest coastal city in Hainan. It has a long history of trading with Vietnam. Although the reputation of Dongfang is not better than Sanya and Haikou, it has a uniquely aboriginal culture and primitive ecological resources. Good products, nature's treasures, rare treasures, different birds and beasts, tambac and rosewood are abundantly produced here. Compared with neighboring cities that are generally over-commercial development, the native resources of oriental ecology have great potential of tourism development.

The project is located on the seashore of Banqiao Town in Dongfang, surrounded by originally ecological tidal flats, with Jianfeng mountain to the east and a wide view towards the western seascape. As a location of five-star resort hotel that is high-quality, the status quo of development of the place that is near the plot is still not mature enough. On the macro level, the nearest airport can only take Sanya or Haikou, which takes more than one hour; Surrounding roads are town-level highways mainly, narrow and old; The level of development of the annex area is low, whose surrounding quality of the supporting is not high. Microscopically, the site is small in size and irregular in shape, surrounded by residential buildings in the plan; The plot is 300 meters away from the sea. Because evacuation of the lake and breakwaters of the slope are obstructed, it is unreachable on foot; The plot ratio of the project is 3.5. Except for 400 guest rooms, the design of about 1,000 sea-view apartments is required. Under these difficulties, if we want to obtain the success of design and management, we will meet the high-level demands of managing companies.

The idea of traditional design obviously does not work. We need to dig deeply in resources such as culture, economy, and landscape and meet the requirements of operation and management in terms of efficiency. More importantly, the unique attractiveness is shaped, which may be the biggest difficulty in this design.

海南东方、西海岸一线海景
原生态人文及景观环境、东南亚知名酒店管理公司、
品质海景、特色水乐园、泰式风情特色
独特的的海南首座水乐园主题亲子酒店
预计开业时间：2020 年

Dongfang, Hainan, the seascape of the west coast line
The environment of originally ecological humanities and landscape, the well-known company of hotel management in Southeast Asia
Quality seascape, distinctive water park, features of the style of Thailand
The first parent-child hotel whose theme is water park with unique personality
Expected opening hours: 2020

31°11′
25.83″ N
121°25′
8.61″ E

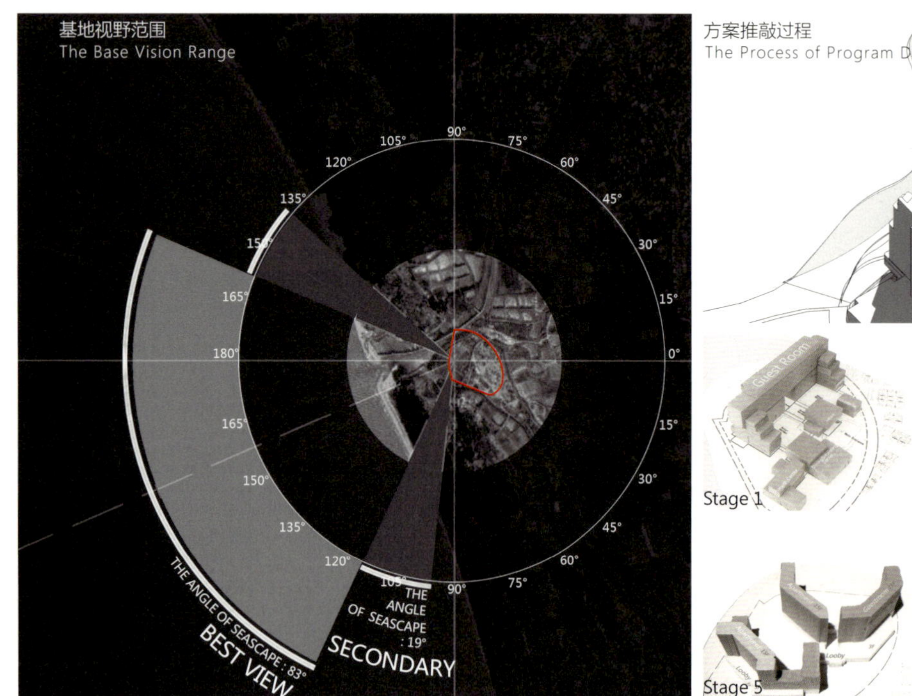

视线分级分析
Sight Classification Analysis

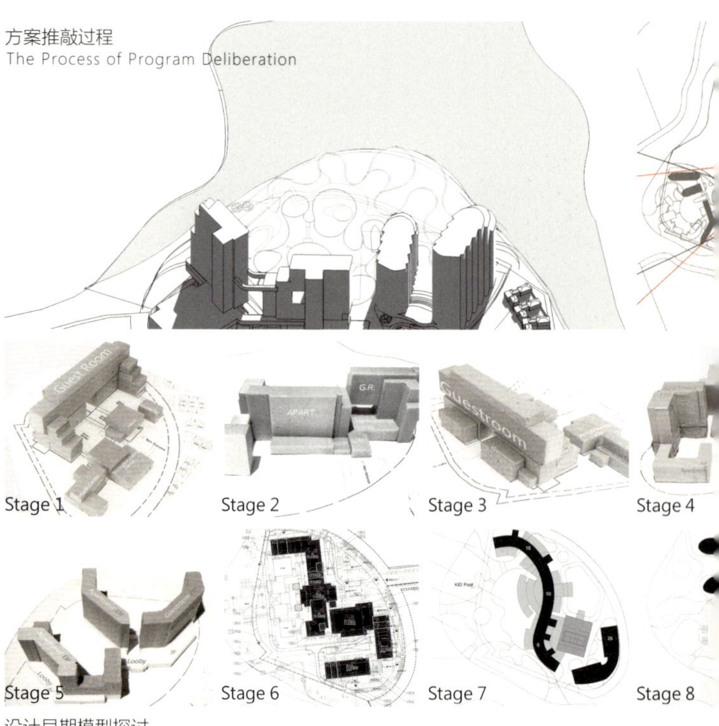

设计早期模型探讨
Discussion on Design Model

“海全景”与“全海景”

地块的核心资源是海景，设计的首要原则就是为所有的住户创造海景视野。依据酒管公司要求，320 间酒店客房必须全正面海景，在地块有限的面海展开面中，如何为 1 000 户公寓创造良好海景视野？设计初始阶段，我们尝试在有限的资源中平衡探讨各体块的组合，探讨规划可行性。

这片地是湾口区的一块扇形滩涂，在边缘处与海之间隔了一道防护林。海景视角约为 120°，其中最佳视角有 83°。经过几轮规划推敲，最终确定将酒店设计为单廊，所有客房正面观海，占据六成海景面，将公寓设计为垂直向海的中廊形式，所有客房斜向观海且避免与酒店的视线冲突，在约四成海景面中实现客房全海景。无论是酒店还是公寓，在阳台上向西望去，都是浩浩荡荡海天一色，夕阳下满眼都是跳跃的金色，蔓延无际。

裙房与景观也融入海的文化元素。从海港特色的水乐园主题景观，到以船为主题的装饰与造型的设计，从海港泊船式的大堂空间设计，到海边餐厅沙滩烧烤等场所营建，设计上都在强化海的印象，拉近人与自然的距离。

"Panorama of Sea" and "the Whole Sea View"

The core resource of the plot is seascape. The primary principle of design is to create a sea view for all residents. According to the requirements of companies of hotel management, There must be wholly frontal seascape in 320 hotel rooms. How can we create a good view of the sea for 1000 apartments in the limited area of the sea? At the initial stage of designing, we try to explore the combination of individual blocks in a limited amount of resources and the feasibility of the programme.

The place is a fan-shaped beach in the bay, separated from the sea by a protective forest at the edge. The visual angle of sea is about 120 degrees and the best visual angle is 83 degrees. After several rounds of planning and scrutiny, we finally decided to design the hotel as a one-sided gallery. All rooms face the sea frontally, occupying 60% of the seascape. The apartment was designed as a vertical corridor to the sea, with all rooms obliquely facing the sea, which avoids the conflict of sight with the hotel. The whole seascape of the rooms is achieved in about 40% of the seascape. Whether it is in a hotel or an apartment, looking westwards on the balcony is a sea of heaven and earth. The sunset is full of jumping gold, spreading endlessly.

The podium and the landscape also incorporate the cultural elements of the sea. From the landscape of water park theme featured in the harbour, to the ship-themed design of decoration and model, from the space design of the harbor berth-style lobby to the construction of restaurants, beach, BBQ and other places by the sea, the designs are all enhancing the impression of the sea, shortening the distance between people and nature.

现状：滩涂用地与山景
Beach Land and Mountain Scenery

感知环境与找寻故事

只有海的资源，这座酒店仍难吸引足够客源。为了找寻契合地域的设计灵感，我们多次实地考察，从三亚一路驱车来到东方。参观完三亚酒店群后，我们一度产生了迷惑，这里的基建及服务都远非现在的东方所能比拟。我们多次探访基地，试图找到地块除了海以外的闪光点。

路还没铺好，黄土上压过深深浅浅的轮印。迎面吹来的是海风的咸味，带着新雨后弥漫的潮气。背后尖峰山微微隆起，脊影沉入积云，掩埋在远处的地平。前面在防护林间几所临时搭建的窝棚，就是这块地里唯一的“房子”。肆意生长的绿如同积水蔓延，灌入沿海滩涂的泥沼中。扒开草丛，水深点的地方甚至能摸出些鱼来。或许是因为雨后的原因，天特别低，又特别宽阔。我们一路走，一路聊，渐渐放开所有思虑，沉浸在游客探索的氛围中。我们聊起儿时的回忆，聊起当年河鲜的美味，聊起在被父母训斥的年纪里痴迷的游戏，聊起不知雾霾是何物时在自然中撒野的心情。渐渐地，我们将眼前的原生态与记忆重叠，一起憧憬着渐渐消失在现代生活中，那些童年时嬉闹的情景。

Perceiving the Environment and Finding Stories

Only the resources of the sea, this hotel is still difficult to attract enough tourists. In order to find the inspiration of design for the region, we visited the site several times and drove from Sanya to the east. After visiting the hotel group in Sanya, we were once confused. The infrastructure and services here are far beyond the modern East. We visited the base many times and tried to find the bright spots of the land except the sea.

The road has not been paved. The deep and shallow wheel marks have been pressed on the loess. The salty taste of the sea breeze is blowing on the face, with the dampness of the new rain. The pinnacle of the mountain rises slightly behind it, and the shadow of the ridge sinks into the cumulus clouds, which is buried in the distant horizon. The tent erected temporarily in the shelter forest is the only "house" in this geographical area. The green that grows like a sprawling water spreads into the shoals along the beach. Opening the grass, you can even find fish in deep water. Perhaps because of the rain, the sky is particularly low and wide. We walked and talked all the way, gradually letting all the thoughts go and immersing ourselves in the atmosphere that tourists explored. We talked about memories of childhood, the delicious taste of the river, the obsessive games in the ages when we are reprimanded by parents and the feelings of playing in the nature without knowing smog. Gradually, we overlap the original ecology with our memory, hoping the scenes of playing in childhood that gradually disappear in modern life.

项目 / Project : 海南金泰湾盛泰澜度假酒店 / Centara Grand Beach Resort Jintai Bay Hainan

类型 / Project Type : 新建度假酒店 / New Construction Resort Type Hotel　　位置 / Location : 中国海南 / Hainan , China

定位 / Project Category : 5 星 / STAR　　修改日期 / Revision Date 2016年3月4日 / 4th March 2016

客房数量 / Total Rooms : 350 间 / Rooms　　修改人 / By : Nattarudee M.

	描述 / Description	面积（平方米）Area (SQ.M)	备注 / Note
1	到达&接待Arrival & Reception		
1.1	到达区 / 雨棚 / 下客区 / Hotel Arrival Area / Porte Cochere / Guest drop off	112	下客区需要有雨棚遮盖，高度至少4米；路宽至少8米（足够3辆车停靠或者同时通过）Guest drop off area shall under roof / cover. Height of roof is at 4 meter(minimum) and width of road is at 8 metre. (enough spaces for 3 cars stopped and/or passing by)
1.2	宴会到达区 / Convention/Ballroom Arrival	112	下客区需要有雨棚遮盖，高度至少4米；路宽至少8米 / Guest drop off area shall under roof / cover. Height of roof is at 4 meter(minimum) and width of road is at 8 meter
1.3	服务台及大堂 / Reception & Lobby	350	层高至少5米，而且建议设置约60平米的团队入住独立接待区域 / Ceiling height shall be at 5 metre. Propose to provide separate group checkin area at 60 sq.m space
1.4	前台 / 工作区 / Front Desk / Work Area	42	建议前台有3个办理入住的柜台 / 团队入住接待区设施2个柜台 / 大堂区域还应设置礼宾部、客户关系、门童等配套 / Propose 3 standing registration counter for Hotel + 2 Standing counter for group checkin. Conciege, Guest Relation and Bell Desk are separate and located in Lobby area
	保险存放室 / Safe Deposit room	28	
	前厅办公室 / FO Office	28	
1.5	行李房 / Luggage Room	42	
1.6	公共洗手间 / Pu (M,F)	84	（男洗手间：5洗手盆 + 5小便池 + 5WC室，女洗手间：6洗手盆 + 6WC室，另外还设置一个残疾人洗手间）Male Toilet 5 basin + 5 Urinal + 5 WC rooms , Female Toilet 6 Basin + 6 WC rooms + 1 Disable toilet
1.7	停车位 / Parking Space		停车位按照当地规范设置；我们建议至少预留3个大巴停车位Number of Carparking space to be comply with local code and recommend adding minimum 3 space for Bus parking.
2	客房 / Accommodation		以350间客房为参照Referred to total number of 350 guestrooms
2.1	双床房 / Double - Double I　227　间 / Rooms	12,712	这227间客房中，包含了79间家庭房（占比35%），配备双床 + 沙发床 / 35% of total 227 Double-Double rooms will be 79 Family rooms with Queen size beds+ sofa beds
	房间套内面积50平米 + 6平米阳台 / Room space at 50 sq.m + Terrace space at 6 sq.m		
2.2	大床房 / Single King Room　105　间 / Rooms	5,880	这105间客房中，包含了37间家庭房（占比35%），配备大床 + 上下铺儿童床供儿童使用；其他68间大床房，考虑配备可收放的沙发床 / 35% of total 105 Single King rooms will be 37 Family rooms with a King size bed plus bunk bed,
	房间套内面积50平米 + 6平米阳台 / Room space at 50 sq.m + Terrace space at 6 sq.m		arrangement for kids. And the other 68 King rooms, to be considered having some/ or all of those
			with pull-out sofa beds.
2.3	套房 / Suite Room (King E　18　间Rooms	1,476	
	房间套内面积72平米 + 10平米阳台 / Room space at 72 sq.m + Terrace space at 10 sq.m		
3	餐饮 / F&B		
3.1	全日餐厅 / Main Restaurant		
	用餐区 / Dinning A　室内 / 半室内 / 室外 / (Indoor/semi outdoor/outdoor)	450	建议设置约400个座位（包括250个配空调的室内座位 + 半开放及开放的座位150个）Propose Reastaurant seating at 400 seats.(Indoor with A/C, 250 seats + semi-outdoor & outdoor space 150 seats
	自助餐区 + 吧台 + 开放厨房 / Buffet Space + Bar + Show Kitchen	120	
	厨房 / Kitchen　(Cooking)	180	约为用餐区&自助餐区 + 吧台 + 开放厨房面积的30% / 30% of total Dining area & Buffet Space + Bar +Show Kitchen
	公共洗手间 / Public (M,F)	84	Male Toilet 5 basins + 5 Urinals + 5 WC rooms , Female Toilet 6 Basins + 6 WC rooms + 1 Disable toilet
3.2	中式特色餐厅 / Spaciality Restaurant 1		
	用餐区 / Dinning A 室内配置空调 / (Indoor, airconditioning)	360	中餐厅建议设置10个包房，约200个座位；包房可以分成小包房和大包房，还可以设置一个连通包房，可以灵活利用；Propose Reastaurant seating at 200 seats as Chinese Specialty Restaurant with 10 Private Dining Rooms
	吧台 / Bar	22	普通包房一般可以坐约12个人 / Required 12 people to fit into private dining room
	厨房 / Kitchen　(Cooking)	108	约占用餐区域的30% / 30% of Dining Area
	公共厕所 / Public T(M,F)	42	（男洗手间：3洗手盆 + 3小便池 + 3WC室，女洗手间：4洗手盆 + 4WC室）Male Toilet 3 basins + 3 Urinals + 3 WC rooms , Female Toilet 4 Basins + 4 WC rooms
3.3	海边餐厅 / Beach Restaurant (THE COAST)		
	用餐区 / Dinning A (Indoor - Outdoor)	400	建议80%的位置为户外，150个位置是户内，约50个位置为半户内 / approximately 80% outdoor. 150 seats as outdoor and 50 seats for semi-indoor
	吧台 / Bar	42	
	厨房 / Kitchen　(Cooking)	180	
	公共厕所 / Public T(M,F)	84	（男洗手间：5洗手盆 + 5小便池 + 5WC室，女洗手间：6洗手盆 + 6WC室，1个残疾人厕所）Male Toilet 5 basins + 5 Urinals + 5 WC rooms , Female Toilet 6 Basins + 6 WC rooms + 1 Disable toilet
3.4	酒廊 / 酒吧 / Lobby Lounge/Bar		建议设置在大堂区域 / Propose location at Lobby Area
	座位区 / Seating Space	100	
	吧台区 / Bar Counter Space	20	
	大堂吧存储空间 / Bar Storage Space	12	
3.5	行政酒廊（高端客人使用）/ The Club (Premium member)		
	酒廊面积 / Room Space	160	室内80平米以及户外80平米 / 80 sq.m. as indoor and another 80 sq.m. for outdoor space
	备餐区 / Pantry　(Cooking)	30	
	公共厕所 / Public T(M,F)	12	（男洗手间：2洗手盆 + 2小便池 + 2WC室，女洗手间：3洗手盆 + 3WC室）Male Toilet 2 basins + 2 Urinals + 2 WC rooms , Female Toilet 2 Basins + 2 WC rooms
3.6	家庭会所（高端家庭客人和小孩一起使用） The Family Club (For Premium member to be used together with their kids)		
	面积 / Room Space	300	
	备餐区 / Pantry　(Cooking)	40	
	公共厕所 / Public T(M,F)	20	（男洗手间：2洗手盆 + 2小便池 + 2WC室，女洗手间：3洗手盆 + 3WC室，3个儿童WC室）Male Toilet 2 basins + 2 Urinals + 2 WC rooms , Female Toilet 2 Basins + 2 WC rooms + 3 WC Kids room
4	宴会及会议设施 / Function Room / Meeting Room		
4.1	大宴会厅 / Ballroom		建议宴会厅高度至少7米而且必须无柱。建议宴会厅可以用隔断墙分隔为两个小宴会厅，灵活使用 / Propose minimum ceiling height at 7 metre and column free inside the room. Recommended to have Operable Wall for possibility to separate one main room into 2 small rooms
	宴会前厅 / Pre Function Space	400	
	宴会厅面积 / Room Space	1000	
4.2	小会议室 / Meeting Room		
	会议前厅 / Pre Function Space	160	
	会议室1 / Meeting room #1	75	
	会议室2 / Meeting room #2	75	
	会议室3 / Meeting room #3	75	
	会议室4 / Meeting room #4	150	
	会议室5 / Meeting room #5	150	
4.3	公共厕所 / Publ (M,F)	150	（男洗手间：10洗手盆 + 10小便池 + 10WC室，女洗手间：14洗手盆 + 14WC室，1个残疾人厕所）Male Toilet 10 basin + 10 Urinal + 10 WC room , Female Toilet 14 Basin + 14 WC room + 1 Disable toilet
4.4	厨房后勤 / Pantry / Kitchen	400	如果可行，建议宴会厅和某些餐厅的厨房共用（设置在相同的位置），以便节省面积）If possibly combine Kitchen to be one between 4.4 and 3.1.3. And that could make kitchen area smaller
4.4	宴会储存室 / Banquet Store	200	

5	水疗及接待处 / SPA & Recreation		
	5.1 水疗 / SPA		5.1 - 5.1.7水疗面积大约为250平米 / SPA area 5.1 - 5.1.7 shall be approximately ± 250 sq.m.
	5.1.1 接待处 / Reception	40	建议在水疗接待处提供一个零售区域以便售卖水疗产品 / Propose provide one retail shop at Spa reception area
	5.1.2 放松区 / Relaxation Area	20	
	5.1.3 茶水间及储存室 / Pantry/Bar Storage Space	12	
	5.1.4 理疗室 / Treatment (6 rooms)	96	
	5.1.5 泰式按摩室 / Thai M (4 rooms)	80	
	5.1.6 脚底按摩室 / Foot Massage	TBD	
	5.1.7 美甲室 / Pedicure + Manicure	TBD	
	5.2 健身中心 / Fitness Centre		
	5.2.1 面积 / Room space	280	
	5.3 水乐园及泳池区 / Water Park and Pool Area		水乐园及泳池区域5.3-5.37的面积大约为2500平米 / Water Park and Pool Area 5.3 - 5.3.7 shall be approximately ± 2,500 sq.m.
	5.3.1 主泳池 / Main Pool	500	建议设置欢乐好玩的水上设备 + 合适的水滑梯 + 以及流转速率合适的懒人漂流河 / Propose Good and Fun Water Play equipments + proper sliders + Good velocity of water in lazy river.
	5.3.2 欢乐池 / 懒人漂流河 / Fun Pool / Lazy River	1000	*建议懒人漂流河的速率约为25 - 30 / 分钟 / Suggested to have velocity at 25-30 metre per minute.
	5.3.2 儿童池 / Kid Pool	100	*建议在儿童池设置2个弧形滑梯，2个不同高度的跳崖 / Suggested to have 2 curved slider in Kids Pool. 2 location for jumping cliff with difference heights.
	5.3.3 泳池栏 / 甲板 / Pool Terrace / Deck	500	*大泳池旁边设置2个长滑梯和3个短滑梯 / 2 Long ride sliders and another 3 short ride sliders at the Main Pool.
	5.3.5 日光浴室 / Sun Lounge Space	500	
	5.3.6 户外淋浴空间 / Outdoor Shower Space	20	
	5.3.7 公共厕所 / Public T (M,F)	60	（男洗手间：4洗手盆 + 4小便池 + 4WC室，女洗手间：5洗手盆 + 5WC室，1个残疾人厕所）Male Toilet 4 basin + 4 Urinal + 4 WC room , Female Toilet 5 Basin + 5 WC room + 1 Disable toilet
	5.4 儿童俱乐部 / Kid Club		
	5.4.1 面积 / Room Space	60	
	5.4.2 厕所 / Toilet	5	
	5..5 大童玩乐区 / 电玩区 / Teen Zoon / E Zone		
	5.5.1 面积 / Room Space	60	
	5.5.2 厕所 / Toilet	5	
	5.7 网球场 / Tennis Court （2 courts）	550	如果户外场地不够，可以考虑设置在宴会厅屋顶；由于这是我们盛泰澜品牌要求，所以至少要有一个网球场 / Possibly on Rooftop (minimium required at least 1 court as this project is Centara Grand)
	5.8 水上运动中心 / Water Sport Centre	100	
6	后勤办公 / Administration		
	6.1 总经理办公室 / GM Office and Meeting Room	24	
	6.2 会议室 / Meeting Room	40	
	6.3 财务办公室 / Finance Office	20	财务办公 + 会计办公 + 采购办公 = 70平米 / Finance Office + Accounting Office + Purchasing Office shall be = 70 sq.m.
	6.4 IT办公室 / IT Office	26	
	6.5 IT控制室 / 服务器室 / IT Control / Server Room	24	
	6.4 会计办公室 / Accounting Office	10	
	6.5 采购办公室 / Purchasing Office	20	
	6.8 收货办公室 / Receiving Office	12	
	6.9 收货平台 / Receiving Dock	20	
	6.10 保安办公室 / Security Office	12	
	6.11 市场营销办公室 / Sale and Marketing Office	30	
	6.12 人力资源办公室 / HR. Office	25	
	6.13 培训室 / Training Room	40	
	6.14 餐饮部办公室 / F&B Office	16	
	6.15 厨师长办公室 / Chef Office	12	
	6.16 管家办公室 / HK. Office	40	
	6.17 总储存室 / Main General Store	100	包括OE等管事部办公室 / included OE , etc. and stewarding office
	6.18 文件储存室 / File & Storage	50	
	6.19 急救室 / First Aid Room	24	
7	员工设施/Employee Facility		
	7.1 总经理公寓 / GM Apartment	TBC	
	7.2 驻点经理公寓 / RM Apartment	TBC	
	7.3 行政管理人员宿舍 / Executive Staff Accomodation	TBC	
	7.4 高级员工宿舍 / Senior Staff Accomodation	TBC	
	7.5 普通员工宿舍 / Standard Staff Accomodation	TBC	
	7.6 员工更衣室 / 洗手间 Staff L (M,F)	220	
	7.8 员工餐厅 / Staff Canteen		
	用餐区 / Dinning Area	200	
	厨房 / Kitchen （Cooking）	60	
8	工程 / 保养 / Engineer / Maintenance		
	8.1 工程 / 保养储存室 Maintenance/ Engineer Storage	84	
	8.2 木匠 / 机电工作室 Carpenter/ M&E Workshop	112	
	8.3 分拣区 Sorting Area	56	
	8.4 总工程师办公室 Chief Engineer Office	20	
	8.5 工程部办公室 Engineer Office	30	
	8.6 机电设备 / 工作间MEP (M&E) Plant room/ Area	TBD	需要机电顾问提供详细要求 / Required MEP (M&E) Engineering Consultant(s)
9	饮料储存室 / F&B Store		
	9.1 干货储存室 Dry Food Store	112	
	9.2 饮料储存室 Beverage Store	56	
	9.3 冷冻室 Freezer Room	56	
	9.4 冷房 Cold Room	56	
10	管家和洗衣 / House Keeping & Laundry		
	10.1 管家储存室 Housekeeping Store	120	
	10.2 化学物品室 Chemical Room	16	
	10.4 洗衣房 Laundry	600	
	10.5 洗衣办公室 Laundry Office	30	
	10.6 布草及制服分配室 Linen + Uniform Distribution	100	
	10.7 布草储存室 Linen Store	100	
	10.8 管家休息室 Maid Station	24	one room on each guestroom floor (with w/c and Ice) 每个客房楼层一间
	10.9 花房 Florist Room	20	
	10.10 干垃圾房 Dry Gabage Room	40	
	10.11 湿垃圾房 Wet Gabage Room	40	

小朋友欢聚水乐园与懒人河
Water Paradise & Lazy River

Sky Garden: Tree House
空中花园：树屋

“船舱”内的 CHECK IN 体验
Check in Experience in Cabin

亲子旅行正呈现趋势
The Trend of Parent-child Travel

感知环境与找寻故事

随着设计展开，我们察觉到，既然移植传统的开发模式可能很难成功，何不另辟蹊径，比如在东方设计一座含五星级酒店的乐园，让它成为整个海南最懂孩子、最值得孩子来玩的场所？管理公司盛泰澜在类似主题酒店的运营上，也有着丰富的经验（泰国芭提雅盛泰澜幻影酒店被誉为最成功的亲子酒店之一，水乐园的成功运营功不可没）。几轮沟通后，我们很快将主题水乐园列为核心设计亮点。此外，主体建筑设计部分也突破了传统的局限，在豪华与舒适之外，融入更多特色的元素，在形态、空间、功能（SPA、餐饮）设计方面均融入泰式文化。

三亚酒店确实成功，但其成功的数量级足够大后，同质化困境随之而来。在本次实践中，尝试以一种新的思路，由乐园亲子主题及异域文化特质稳固产品基础，定制一种特色的酒店模式。这不只是在设计一座酒店建筑，也是在探寻一种让家庭和孩子都深感轻松愉悦的度假方式。

Perceiving the Environment and Finding Stories

As the design unfolded, we noticed that since it may be difficult to succeed in transplanting the traditional development model, why not find new ways, such as designing a paradise with a five-star hotel in the east, making it become the place where children are understood most and where children play meaningfully? The management company Shengtailan also has extensive experience in the operation of similar themed hotels (Santa Phantom Mirage Hotel, Pattaya, Thailand is regarded as one of the most successful parent-child hotels, the success of the water park contributing to its success). After several rounds of communication, we quickly listed the theme of water park as the core design lightspot. In addition, the main part of architectural design also broke through the traditional limitations. In addition to luxury and comfort, it incorporated more elements of special features. in the . The Thailand culture is integrated into the design of the form, space and function (SPA, catering).

Sanya Hotel is truly successful. However, after its success is large enough, the homogeneity dilemma comes along. In this practice, we try to use a new idea to stabilize the base of product and establish a distinctive hotel model according to the parent-child theme of the park and the characteristics of exotic culture. This is not just about designing a hotel building, but also exploring a way of vacation to make the family and children feel relaxed and happy.

到达体验：船与木
Arrival Experience: Ship with Wood

泰餐特色美食
Food Attraction·

一间制造亲子旅行话题的酒店
A Hotel Makes the Topic of Parent-child Travel

大堂吧与“船”
Lobby Bar and "Ship"

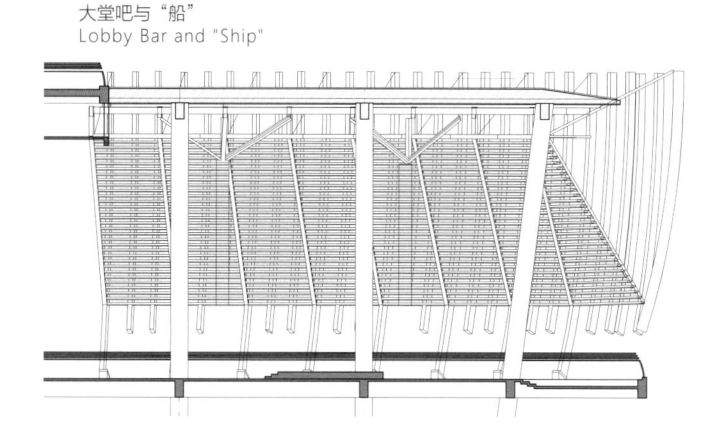

Lifestyle +Funny Trend

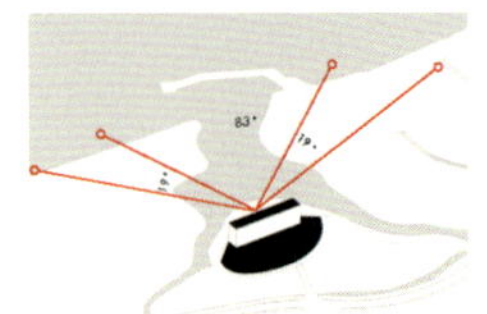

长 152m 层数 12
length 152 米 floors 12

一 字形
Shape 一

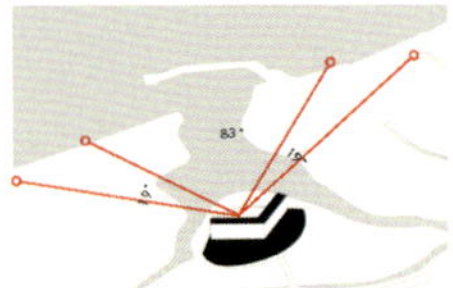

长 182m 层数 10
length 182 米 floors 10

V 字形
Shape V

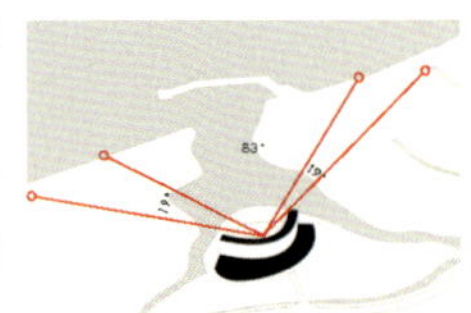

长 203m 层数 9
length 203 米 floors 9

C 字形
Shape C

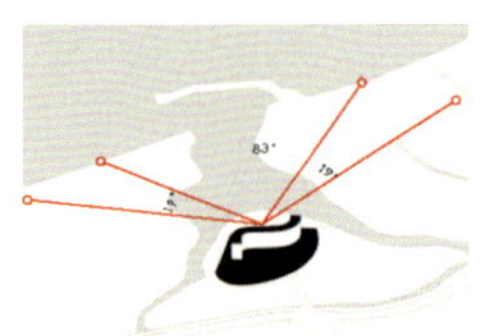

长 152m 层数 12
length 152 米 floors 12

S 字形
Shape S

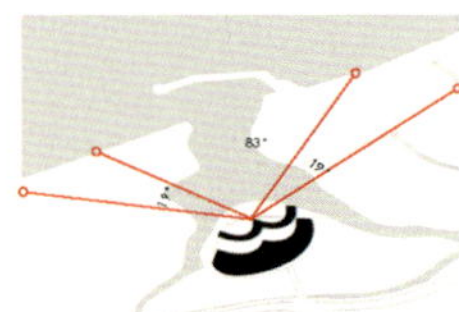

长 120m 层数 8
length 120 米 floors 8

M 字形
Shape M

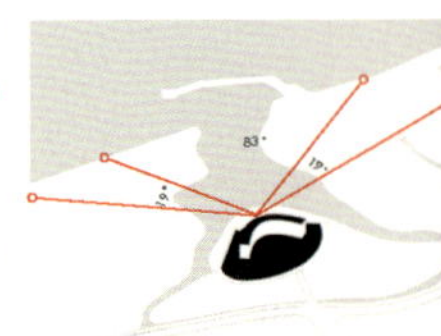

长 203m 层数 9
length 203 米 floors 9

C 字形
Shape C

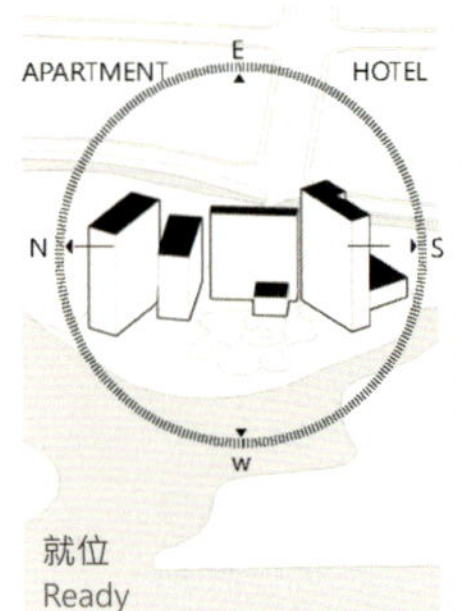

就位
Ready

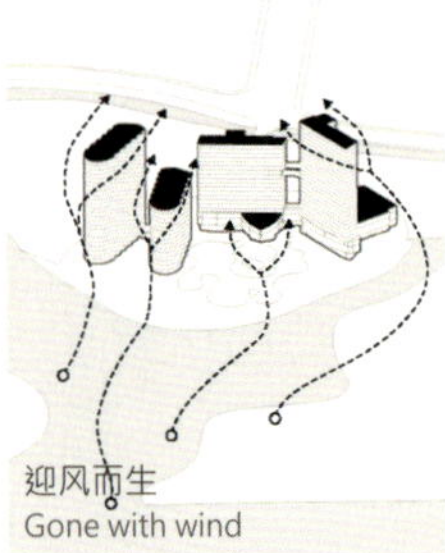

迎风而生
Gone with wind

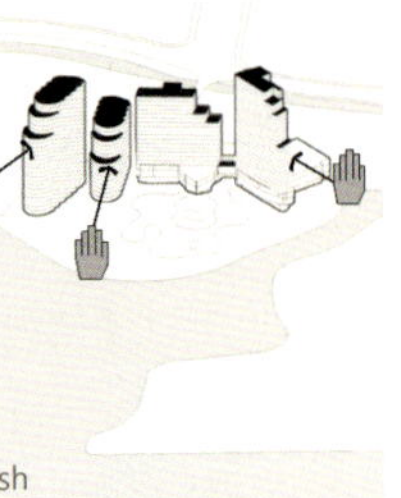

推
Push

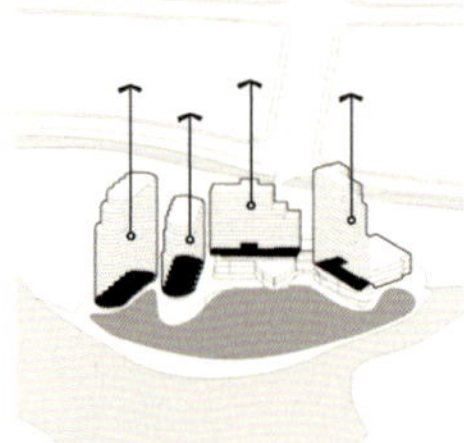

漂浮
Float

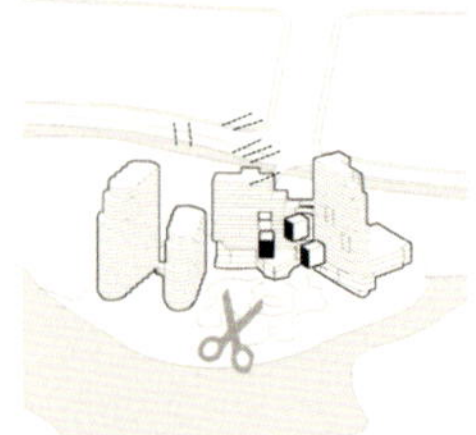

Cut
裁剪

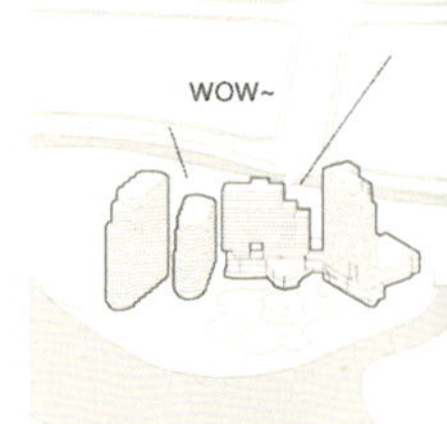

完成
Finish

“东方船说”的故事

基于港口城市、海边地块、泰式特色三个背景特点，经反复考虑，设计选择将港、船、木、水作为基本元素，构思整体设想：规划中将建筑组合成向海进发的动势，与水乐园组成一座“前进”的港湾；建筑的端部形体基于风的考虑，轮廓设计为船头形，提供端部客房更好的视野；大堂的设计灵感取自“黎族茅草船形屋[2]”与“郑和宝船[3]”，是一个大木构组成的船舱空间模块，让游客由“登船”开始拜访之旅；建筑的主色彩由白色与木色组合，塔楼的弧形金属板在阳光下渲染出游轮般的质感；海边餐厅位于主泳池与“Lazy river[4]”之间，在倒扣的船体屋顶下，木色与环境自然融合，露天酒吧与 BBQ 的热闹人气，渲染了轻松的海边度假氛围 ...... 我们尝试将自然、文化、地域等多方面要素提炼定制，用特色的设计语言，以孩子能欣然读懂的方式描绘出来，让这里成为最东方最欢乐的酒店。

“设计中我们研究了几乎所有可能存在的形体模式（上图），最终从‘C’形推演确定了‘环抱式’，强化海景视野的平衡创造。”

The Story of "the Saying of Oriental Boat"

Based on the three background characteristics of the port city, seaside plots and Thailand characteristics, the design chooses port, ship, wood and water as the basic elements, conceiving the overall idea: the plan will combine the building into a trend of moving toward the sea, forming a "forward" bay with the water park ; Considering wind, the contour of the end of the building is bow-shaped, providing a better view of the end rooms; The inspiration of lobby design comes from the "Thatched Boat- shaped housing of Li Clan[2]" and "The Zheng He Treasure Boat[3]". It is a cabin module consisted of a large wooden structure, allowing visitors to start a tour from the "boarding"; The main color of the building is composed of white and wood. The tower's curved metal plate is cruise-like texture in the sun; The seaside restaurant is located between the main swimming pool and the "Lazy river[4]". Under the inverted hull roof, the wood color is naturally merged into the environment. The popularity of open bar and BBQ render a relaxed atmosphere of seaside resort and so on. We try to abstract and customize many elements of nature, culture, and region, using a distinctive design language to depict them in a way that children can readily understand in order to make it the most joyous hotel in the east.

"In the design, we studied almost all possible physical patterns (above), and finally determine 'encirclement' deducted from the 'C' model, strengthening the balanced creation of sea vision."

模型："港湾"与演绎自郑和"宝船"的大堂吧造型
Model: "Harbour" and Interpretation of the Lobby Bar from Zheng He "Treasure Ship"

"绘本之外"的故事

看不见的价值往往更能体现专业度，正因为高星级酒店的运营复杂，与管理公司的配合变得至关重要。在最初的设计任务要求中，酒管列出了百条功能要求，随后的补充甚至有功能关系在内的各种分析图，这是很认真的专业度。后来展开的工作中，平面设计从各业态的布局到每个房间的大小和形状、开门，从流线的组织到每个节点的梳理，都经历了多轮打磨[5]。

亲子型酒店有不少特色功能定制，例如符合儿童活动习惯的客房、"Family Lounge"、"E-zone"、"Ice-bar"、"Check-in"区域空间，都有不同的理解。另外，从运营思路出发的设计方式也让我们受益匪浅，例如按我们按传统理解，SPA 作为 Centara 独到的特色功能之一，曾屡获殊荣，如放在景观视野较好的区域将品质更高，而酒管公司则认为 SPA 属目的性消费，满足安静与私密已足够，有限资源应优先让给非定向型的经营功能。

Stories of "Beyond the Picture Book"

Invisible values often reflect professionalism. Because the operation of high-star hotels is complex, it's crucial to cooperate with management companies. In the initial design tasks, the hotel management listed hundreds of requirements of function, followed by various analysis charts including functional relationships, which is a serious degree of professionalism. In later work, graphic designs are experienced many rounds of polish from the layout of various formats to the size, shape and opening doors of each room, from the organization of streamlines to the grooming of each node[5].

Parent-child hotels have a lot of features that have different understandings to customize, such as rooms that meet children's habits of activity, Family Lounge, E-zone, Ice-bar, regional space of Checkin. In addition, we also benefited from the approach of design based on the operational thinking. For example, according to our traditional understanding, SPA has won awards as one of the unique features of Cerrara. The region with good view of landscape is in a higher quality. Hotel management think that the SPA belongs to purposeful consumption, meeting the quietness and privacy, which is enough. Limited resources should give priority to non-oriented business functions.

01. 大堂吧 The Lobby Bar
02. 文化展示 Culture Show
03. 大堂 Lobby
04 前台 Front Office
05. 行李间 Luggage
06. 预订部 Reservations
07. 礼宾部 Concierge
08. 下沉庭院 Sinking Courtyard
09. 大堂等候区 Lobby Waiting Area
10. 冰淇淋店 Ice Cream
11. 儿童步径 Children Path
12. 精品店 Boutique
13. 餐饮接待 Reception Area
14. 中式餐厅 Chinese Restaurant
15. 包间 PDR
16. 露台 Balcony
17. 备餐 Prepare
18. 洗碗区 Dishwashing Area
19. 厨房 Kitchen Area
20. 餐厅临时储藏 Storage Room
21. 团队到达 Group Check in
22. 宴会厅 Ballroom Hall
23. 衣帽间 Clook Room
24. 收纳间 Pocket Room
25. 视频控制室 AV Room
26. 宴会厅 Ballroom
27. 精品店 Banquet Store
28. 贵宾室 Bridal Room/VIP
29. 电梯厅 Elevator Lobby
30. 餐厅和会议入口 Diner&Meeting Entrance
31. 落客区 Drop off Area

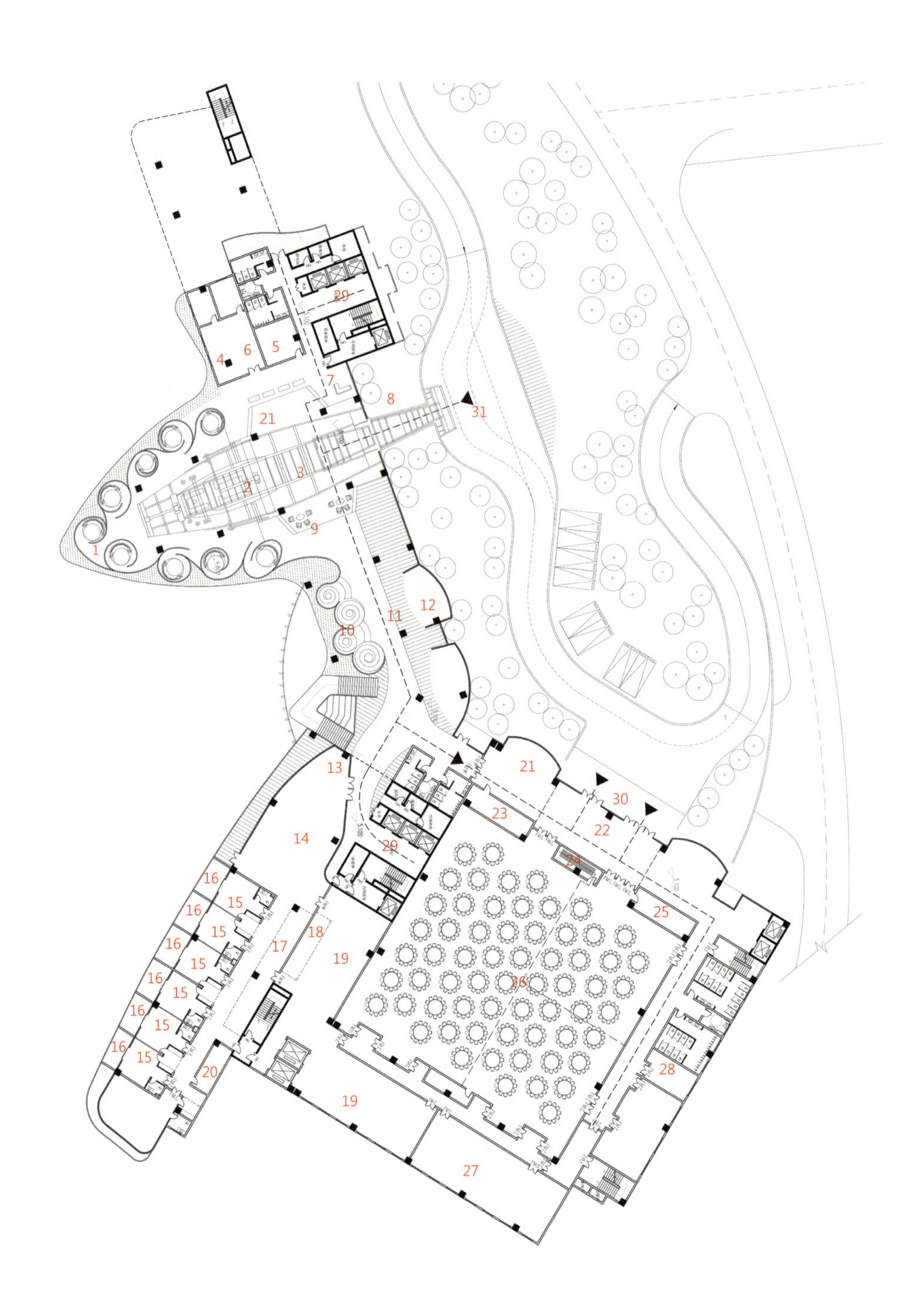

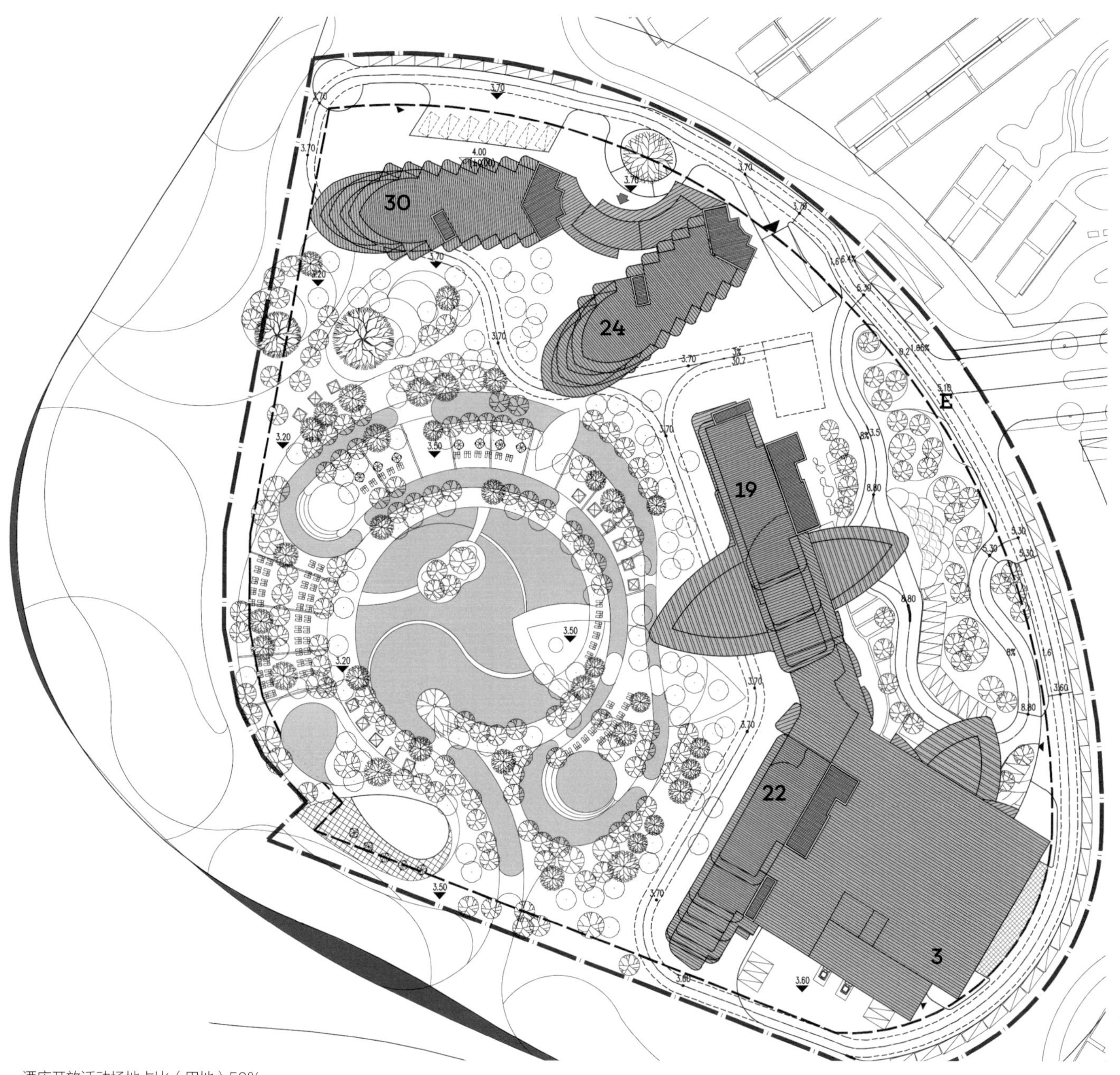

酒店开放活动场地占比（用地）50%
The Open Area of the Hotel Occupies 50% of the Land

中庭视野（邻交通核 A）
Atrium Vision

大堂吧空间推敲过程
Process of Lobby Bar Space Design

“绘本之外”的故事

酒店设计的专业度体现在各处细节中，例如在宴会厅的设计中，不仅需关注其各流线关系、配送方式及空间格局，还需结合未来功能转换的可能，调整平面形状，而本案则根据特殊业态的需要设计为方型；SPA 的位置临近儿童娱乐区，这样既可以托管小朋友又可以体验康体；结合当地气候，我们设计了全开敞的酒店环境，这对于流线组织和场景营建、机电设计都提出了更高的要求，而由此节省的暖通和照明成本也为这座酒店的运营减小了压力。

类似设计过程中的细节，不胜枚举。但最终正是这些细致而周到的考量，才让故事不只是童话，让设计有资格对话现实。

Stories of "Beyond the Picture Book"

The professionalism of the hotel design is reflected in various details. For example, in the design of the banquet hall, the streamline relationships, distribution methods, and spatial patterns need to be concerned and the shape of plane need adjusting according to the possibility of functional conversion in the future . According to the needs of the special format, it is designed as a square shape; The location of the spa is close to the children's area of entertainment so that children can be cared for and the health can be experienced; Combined with the local climate, we designed a fully open environment of hotel, which has put forward higher requirements for streamline organization, scene construction and electromechanical design. And the resulting savings in costs of HVAC and lighting have also reduced the pressure on the operation of the hotel.

Similar details in the process of design are numerous. But in the end, it was these meticulous and thoughtful considerations that made the story more than just a fairy tale, the design being qualified to have a dialogue with reality.

海边餐厅、公寓及酒店塔楼推敲模型
The Scene after the Scaffold is Dismantled

注：

1 盛泰澜酒店及度假村集团（Centara Hotels & Resorts）为泰国最大的酒店管理公司之一，其业务分布于世界各地的重点城市及度假热点，尤其在印度洋和东南亚地区影响力较大。

2 茅草船形屋，是黎族最为典型的建筑。从外形上来看像一艘倒扣着的船，一般呈长方形，以竹子和木头用于搭建屋子构架，用藤条固定，以淤泥筑墙，干草遮盖，冬暖夏凉。海南岛保存船型屋最为完好的 3 个村庄均位于东方，其营造技艺也于 2008 年被列入我国国家级非物质文化遗产名录。

3 郑和宝船，是郑和下西洋船队中最大的海船“长四十四丈，阔一十八丈。”是当时世界上最大的木制帆船，设计参考 2014 年宝船结构模型，以主龙骨与几十根肋骨搭建的方式，完成酒店入口“船”空间设计。

4 Lazy river，是盛泰澜在亲子酒店水乐园中的特色项目，以自然坡差结合丛林、河流、瀑布营造的漂流体验。

5 以上内容主要为建筑方案深度，室内、景观等顾问单位尚未介入。

Note:

1 Centara Hotel & Resorts is one of the largest companies of hotel management in Thailand. Its business is located in key cities and resort around the world, especially in the Indian Ocean and Southeast Asia.

2 Thatched boat-shaped houses are the most typical buildings of the Li nationality. From an appearance point of view, it looks like an upside-down ship, generally rectangular, with bamboo and wood used to build the frame of house, fixed with rattan, building walls with sludge, covered with hay, which is warm in winter and cool in summer. The three most well-preserved boat houses in Hainan Island are located in the east, whose building skills were also listed in Chinese national intangible cultural heritage in 2008.

3 Zheng He Treasure Boat is the largest seabed in Zheng He's Western Fleet. The length is 44 zhang and the width is 18 zhang. It was the largest wooden sailboat in the world at that time. The design was based on the structure model of the treasure ship in 2014. The main keel and dozens of ribs were used to complete the design of "ship" space in the entrance of hotel.

4 Lazy river, a special project of the parent-child hotel water park of Shengtaile , creating a experience of drifting experience by natural slope difference combined with jungle, rivers and waterfalls.

5 The above content is mainly the depth of the construction plan. The consultants such as the interior and landscape have not yet stepped in.

手工模型
Manual Model

龚兵 / 马如柏 / 王涌臣 / 金鹏
Bing Gong / Rubai Ma / Yongchen Wang / Peng Jin

松木板 /PVC/ 瓦纹纸
1 : 1000

水乐园与“环抱式”规划设计
Water Park and "Encirclement" Planning and Design

木模型：酒店塔楼
Wooden Model: Hotel Tower

木模型：公寓塔楼
Wooden Model: Apartment Tower

酒店入口：船与木的元素演绎
An Element of Water Park, Boat and Wood

2016 年 7 月，摄于泰国芭提雅盛泰澜酒店（2007 年开业），独创水乐园主题酒店
Photographed at Centara, Pattaya, Thailand (Opened in 2007), Original Water Park Theme Hotel, 2016,7

看起来不像，其实就是 / 兰州市第一人民医院综合楼

Not Look Like, It's Actually / The Comprehensive Business Building of the First People's Hospital of Lanzhou

兰州

占地面积　：2 168 平方米

建筑面积　：46 276 平方米

建筑高度　：81 米

设计时间　：2016

Lanzhou

Floor Space : 2,168 m^2

Gross Floor Area : 46,276 m^2

Building Height : 81 m

The Time of Design : 2016

充满活力的新医疗空间
Vibrant New Medical Space

“机场比婚礼殿堂见证了更多真诚的吻，医院比教堂听到了更多的祈祷。”

——Shreya Ayanna Chaudhary

医院，是绝大多数人与这个世界相遇的起点与相别的终点，这是个特别的场所，生和死、欢和悲、软弱与坚强、残忍与希望都在这里汇集。

这是省会城市三甲医院的综合楼，虽然本次设计的内容是空间梳理与造型设计，但带着对场所特殊的尊敬与理解，让空间释人以温暖、亲和，让建筑体现人性关爱的感受是设计优先考虑的问题。

设计之初，我们调研了近年建成的优秀医院案例。发现在维持高效率的前提下，将医院的空间设计的更自由、更宜人、更充满阳光是普遍趋势，将建筑形象做得更生态、更轻盈、更时尚简洁也已是潮流——澳大利亚早在 2013 年建成的奇伦托夫人儿童医院就已将“去医院化的心理感受”作为核心设计出发点之一。而通过实地调研发现，兰州当地的医院普遍老旧、严肃而封闭，在入手本项目后，我们尝试通过思考理念的革新设计，以用户心理体验为出发点，挖掘周边环境资源优势，既简单又特别的形式，提升形象审美，以阳光与轻盈的“白”重塑医院形象。

"The airport has witnessed more sincere embrace than the wedding hall. The hospital heard more prayers than the church."

——Shreya Ayanna Chaudhary

The hospital is the starting point of meeting the world and end point of leaving the world for most people. This is a special place where birth and death, joy and grief, weakness and strength, cruelty and hope all come together.

This is the comprehensive building of the top-three hospital of provincial capital city. Although the content of this design is spatial grooming and model design, it has a special respect and understanding to the venue, allowing the space to release people with warmth, affinity, and letting the building reflect the feeling of human care, which the priority of design.

At the beginning of the design, we investigated the outstanding hospital cases built in recent years. It is found that under the premise of maintaining high efficiency, it is a general trend to design the space of the hospital more freely, more pleasantly, and more sunny. It is also a trend to make the building image more ecological, lighter, and more stylish and concise. The Madame Cilento Children's Hospital, built in 2013, has already considered the "psychological impression of going hospital" as one of the starting points of the core design. Through on-the-spot investigation, we found that local hospitals in Lanzhou are generally old, serious, and closed. After starting this project, we tried to use the innovative design of thinking concepts to take advantage of the user's psychological experience as a starting point to tap the advantages of surrounding environmental resources. It is simple and special. In the form of image, it enhances the aesthetic of the image and reshapes the image of the hospital with the "whiteness" of light and lightness.

省会城市、重点医院、医技综合楼
与旧院区的对比与融合、从用户心理出发的大胆设计
契合功能与地域的创新研究

Provincial Capital City、Key Hospital、Complex Building
The contrast and integration with the old hospital area
and the bold design from the user's psychology
research on the innovation of function and region

手工模型
Manual Model

龚兵 / 马如柏 /
Bing Gong / Rubai Ma

PVC
1 : 1000

36°04′
34.31″ N
103°46′
9.48″ E

视野关系分析
Analysis of the Visual Relationship

底层效果：韵律的变化
Effects of the Bottom: Changes in Rhythm

阳光的苗床

虽在西北，但这里并不是灰蒙蒙一片。新楼与主楼间的院子里绿植密布，站在塔楼上能越过街区一直看到远山。而提到医院，大多人头脑中的图景都只有压抑——病痛的呻吟、机械的叫号、排队的烦躁。离开病房，走廊尽头心理感受上也是灰蒙蒙的窗，透过些微弱的光，一切都在封闭的环境中循环，似乎只有愤怒的争执能让心情打开一些口子。我们希望打破这种冰冷的环境，引入这些环绕在医院旁的景致与阳光。

在空间上，我们优化了中庭的布置。掀开顶板，加上玻璃天窗，并在中庭及天窗的四周种上绿植，让阳光植入每个人的视野。调节中庭形状，以错动的体块拉伸空间，使其在视觉比例的调节中更舒展开放。改善中庭附近的交通核布置，让流线通畅的环绕中庭，让空间进一步收放自如。阳光将所有人都联系到中庭，配以简洁明快的室内设计，让这里成为整个医院的呼吸器。

The Seedbed of Sunshine

Although it is in the northwest, it is not grey. The yard between the new building and the main building is covered with greenery. Standing on the tower can see from the block to the distant mountains. When it comes to hospitals, most people's minds are suppressed only by the moan of sickness, mechanical sound, and irritability in the queue. Leaving the ward, at the end of the corridor, there was a grey window on the psychological sensation. Through some faint light, everything circulated in a closed environment. It seemed that only an angry dispute could make the mood open up. We hope to break the icy environment and hope to introduce these views and sunshine that surround the hospital.

In space, we have optimized the layout of the atrium. Opening the top plate, adding the glass sunroof, and planting green plants around the atrium and skylights make the sun implant in everyone's view. We adjust the shape of the atrium, stretching the space with changing body mass, which makes it more open in the adjustment of the visual proportion. We Improve the traffic and layout near the atrium so that the streamline can smoothly flow around the atrium, allowing the space to move more freely. The sun connects everyone to the atrium, with a concise interior design, which makes it become a respirator throughout the hospital.

沿街效果图

The Effect Picture along the Street and Material Joint Model

会跳舞的白

白，几乎是现代建筑的代名词，因为它能去除繁冗的装饰，以简洁的审美重塑造型与空间的联系，以光影的表演诱发行为与时空的对话。同时，白也是医院的属性——卫生、高效与冷静。然而时过境迁，白也不再只是冰冷与静止的颜色，有了更多层次、动作与表情。

我们选用了白色杆件作为造型最重要的元素，用杆件的粗细进退、肌理的起伏深浅，表现白更多的光影魅力。在裙房扬起的端部，杆件不仅遮住背后楼梯间，而且以上扬的造型摆脱了重力的束缚。在塔楼方正的轮廓底部，不仅以缩进的阴影飘浮起岩石般的体量，而且以无约束的脊线收束上下造型节奏的对比，在自由中表达出白的轻盈。

The White That Can Dance

White, is almost synonymous with modern architecture. Because it can remove the tedious decoration, reshaping the relationship with the space by a simple aesthetics, luring a dialogue between time and space by the performance of light and shadow. At the same time, white is also the property of the hospital - health, efficiency and calmness. However, times have passed and circumstances have changed, white is no longer just a cold and static color. It has more levels, actions and expressions.

We chose the white bar as the most important element of the model. We used the thickness of the bar to advance and retreat, and the undulation of the texture to express more charm of light and shadow. At the raised end of the podium, the poles not only cover the stairwells behind them, but also break away from the shackle of gravity by the bullish model. At the bottom of the outline of tower's square, not only does the rock-like volume float by the shadow of indentation, but also the unconstrained ridgeline contrasts the rhythm of the upper and lower styling, expressing the lightness of the white in freedom.

医院裙房 3F，院落中树叶茂密，越过街道可观远山
The Courtyard is Full of Leaves in Podium 3F of the Hospital. Forane Mountains Can be Watched Cross the Street

塔楼 12F，视线可以穿越高楼林立的天际线，看到远山起伏
In Tower 12F, the Sight Can Cross the Skyline of a Tall Building, Seeing ups and downs of Mountains.

看则约束・实则成全

设计不是单纯的模拟和复制，而要表达主题意志。而在现实条件的约束下，有时不得不采用一些特殊做法。

例如由于医院建筑不被允许使用幕墙，所以为了表达出设计中轻盈自由的造型，我们必须用窗墙系统模拟出幕墙的效果。经过多方案比较，我们决定将外立面白色杆件直接嵌入墙体，窗框部分的结构作用保留，外部则是成品一体的装饰杆件。起初我们担心会影响采光、开窗和清洁，但实际安装上后我们发现不仅使用没有问题，而且有了意外框取景深的效果。并且由于杆件一体化设计，节能方面也有较好表现。

It Looks Like Constraints. It's Actually Help.

Design is not a simple simulation and replication, but the expression of the theme will. Under the constraints of actual conditions, sometimes there are some special practices that have to be adopted.

For example, because the hospital building is not allowed to use curtain walls, we must use the system of window wall to simulate the effect of the curtain wall in order to present the model of lightness and freedom of the design. After comparing multiple plans, we decided to directly embed the white member bar of the facades into the walls. The structure of the window frames remains, while the outer ones are finished rods of decoration. At first, we feared that it would affect lighting, windows and cleanliness. However, after actual installation, we found that not only was there no problem in use, but also there was an unexpected effect of deciding the depth of field. And due to the integrated design of the rods, there is also a good performance in terms of energy saving.

接近结构封顶层，建造过程
Close to the Top of the Structure,
the Construction Process

在塔楼的材质选用上，我们原本希望用铝板，但结合造价考虑最终用了水泥板材质。经过设计，三种灰度的水泥板以竖向有机分缝（模度 300mm 最宽 1 200mm），与规则阵列的杆件共同形成有趣的立面节奏。而水泥板介于石材与涂料间的材质效果，也让这栋建筑有了既自然又精美的质感。

无论是空间的营造，还是造型的表达，我们都试图突破传统医院的样本，以温暖而轻盈的氛围体现人文关怀，以时尚而雅致的品位赋予生命更高崇的尊严。医院原本就应是冬夜里的灯笼，传递更多的安心。

In the material selection of the tower, we originally wanted to use aluminum panels. But considering the cost, we finally used the material of cement board. After the design, the three gray-scale cement slabs have vertical organic joints (the width is 300mm and the maximum is 1200mm), which forms an interesting rhythm of facade with the regular array of bars. The material effect of the cement board between the stone and the paint also gives the building a natural and exquisite texture.

Whether it is the creation of space or the expression of styling, we are trying to break through the samples of traditional hospitals, reflecting the humanistic care by a warm and light atmosphere, giving life a higher dignity by a stylish and elegant taste. The hospital should have been a lantern on a winter night that conveys more ease.

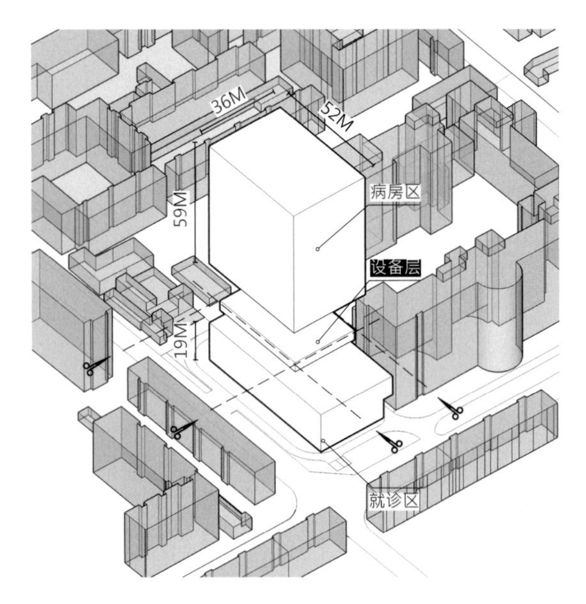

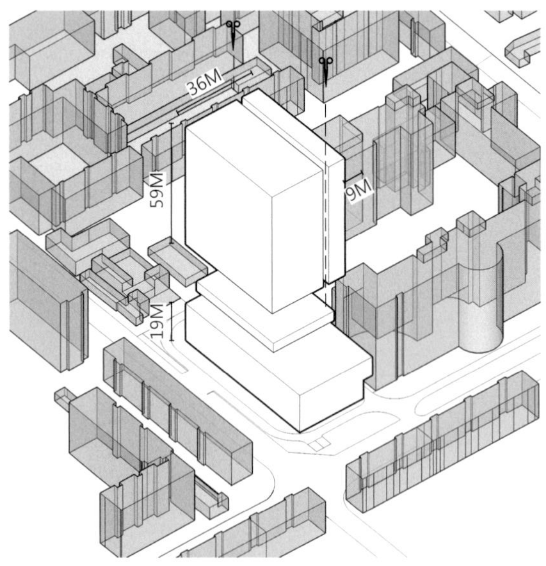

二段式处理

Two-stage Processing

通过设备层将塔楼和裙房分离，实现病房区和就诊区的互不干扰。

The tower and the skirt room are separated by the equipment layer to realize the non-interference of the ward area and the treatment area.

塔楼体块划分

Tower Block Division

面主形象界面将塔楼体块划分为两部分，

强化体量的挺拔感。

The surface main image interface divides the tower body block into two parts, strengthens the body quantity the upright feeling.

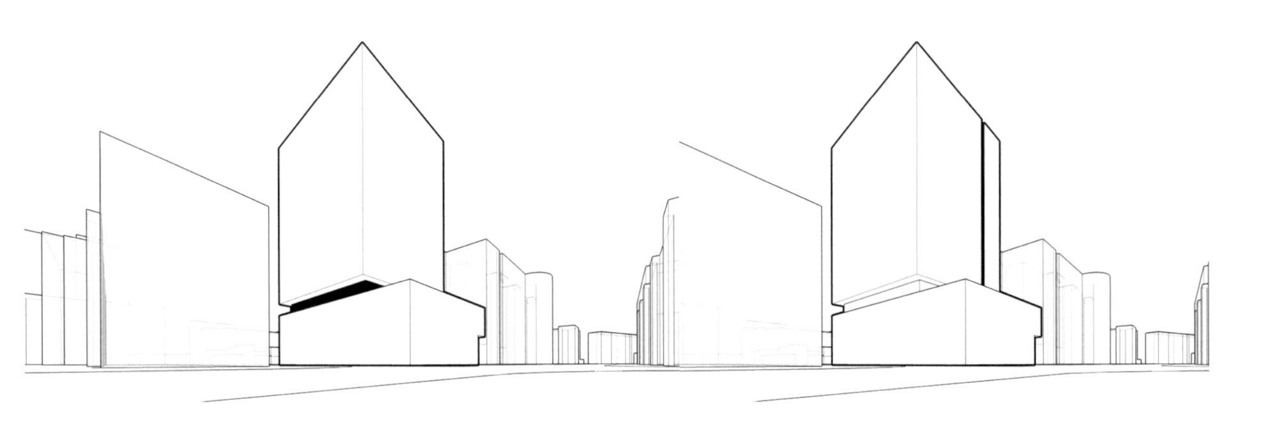

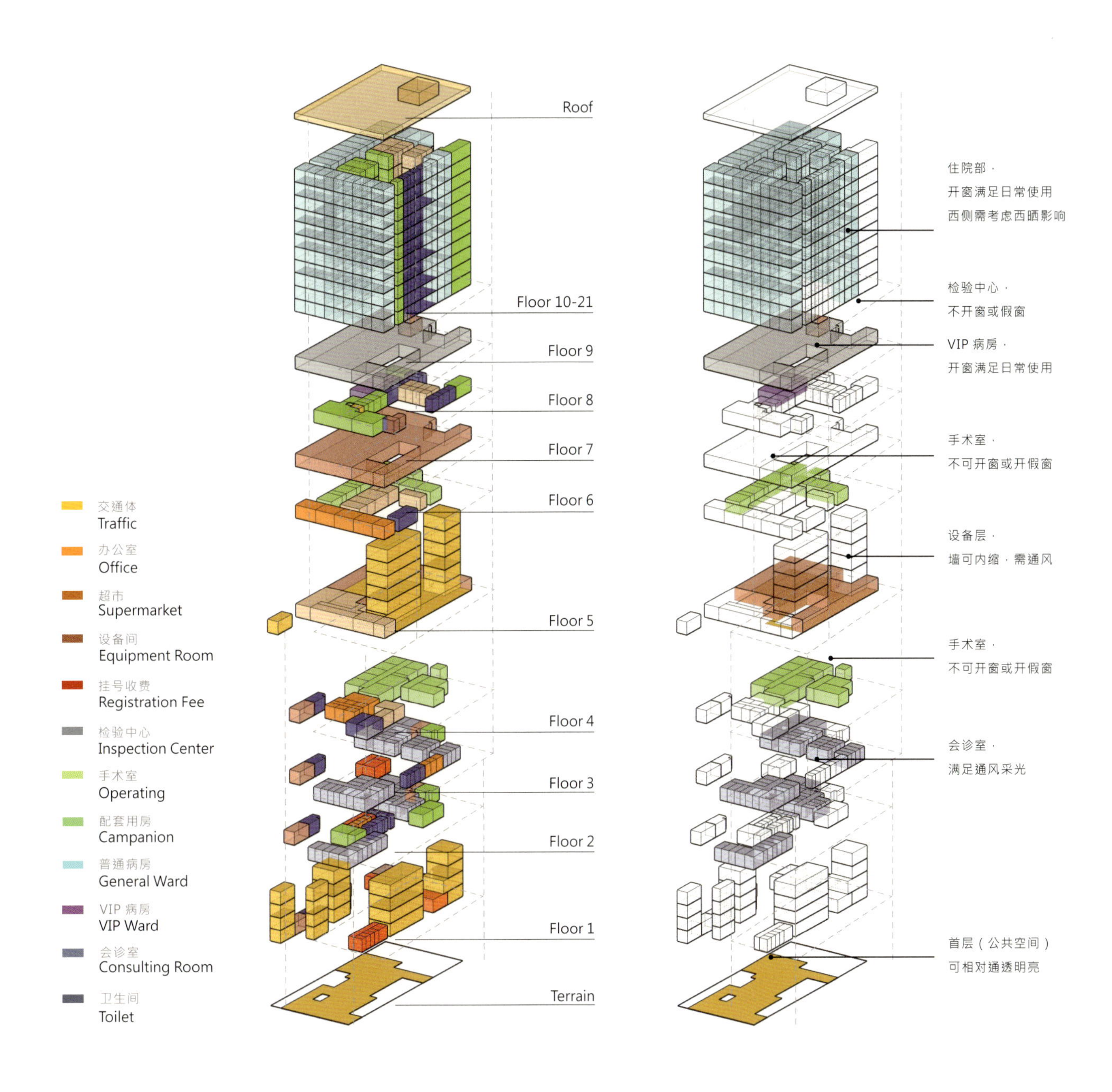

Roof
Floor 10-21
Floor 9
Floor 8
Floor 7
Floor 6
Floor 5
Floor 4
Floor 3
Floor 2
Floor 1
Terrain
交通体
Traffic
办公室
Office
超市
Supermarket
设备间
Equipment Room
挂号收费
Registration Fee
检验中心
Inspection Center
手术室
Operating
配套用房
Campanion
普通病房
General Ward
VIP 病房
VIP Ward
会诊室
Consulting Room
卫生间
Toilet
住院部，
开窗满足日常使用
西侧需考虑西晒影响
检验中心，
不开窗或假窗
VIP 病房，
开窗满足日常使用
手术室，
不可开窗或开假窗
设备层，
墙可内缩，需通风
手术室，
不可开窗或开假窗
会诊室，
满足通风采光
首层（公共空间）
可相对通透明亮

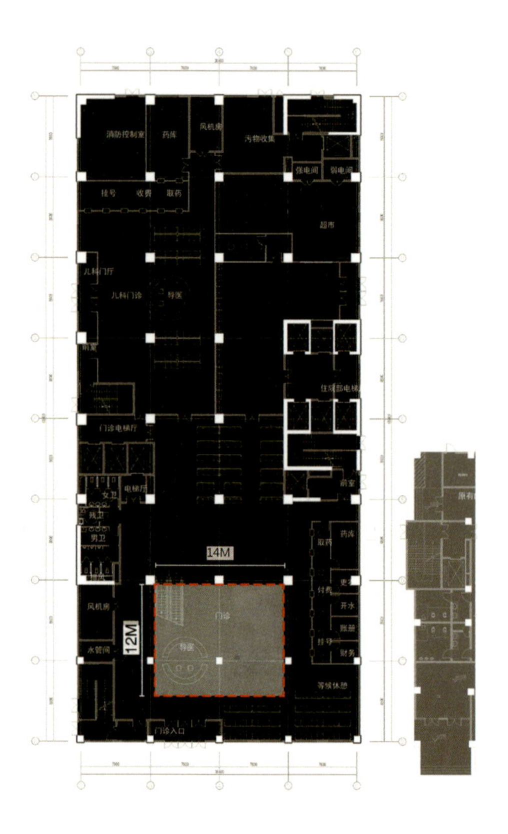

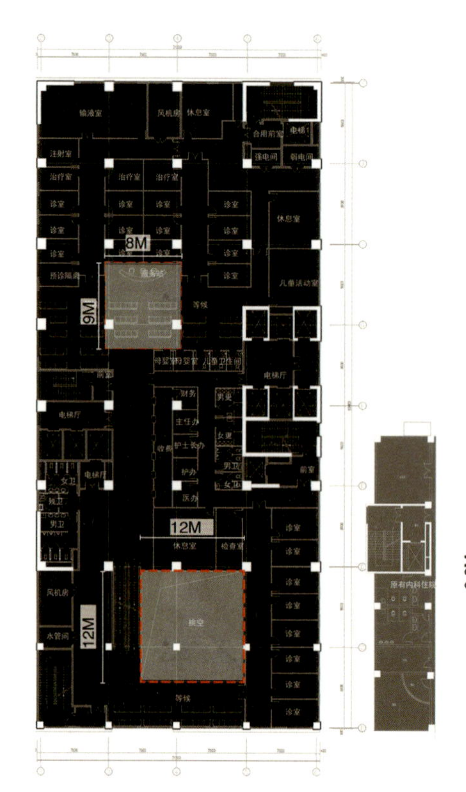

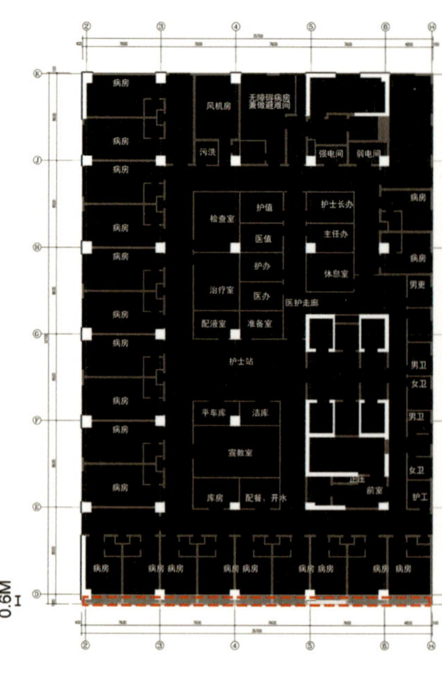

Plan 1

设置通高门诊大厅

Set Up an Outpatient Hall with High Access

原始通高投影面积　: 14×8=112 m²

优化后通高投影面积 : 14×12=168 m²

Original High-Flux Projection Area: 14x8=112 m²

Optimized Rear-pass High Projection Area: 14x12=168 m²

On the First Floor
Enhance the Sense of Experience
Construction Area Unchanged

Plan 2

设置通高儿童门诊

Set Up an Outpatient Hall with High Access

原始通高投影面积　: 0×0=0 m²

优化后通高投影面积 : 9×8=72 m²

Original High-Flux Projection Area: 0x0=0 m²

Optimized Rear-pass High Projection Area: 9x8=72 m²

增加主中庭挑空面积

Increase the Empty Area of the Main Atrium

原始挑空面积　: 12×8=96 m²

优化后挑空面积 : 12×12=144 m²

Original Empty Area: 12x8=96 m²

Optimized Empty Area: 12x12=144 m²

On the Second Floor
48m - reduce Construction Area
Consulting Room Reduced 5

Plan 10-21

增加南侧病房面积

Increase the Ward Area on the South Side

优化后南向病房增加面积 : 0.6×35.7=19.0 m²

Optimized Increasing
Area of South Ward: 0.6x35.7=19.0m²

On the Tenth to Twenty-first Floor
Monolayer Area Increased 19.0m2
The Total Area Increased 285.0m2

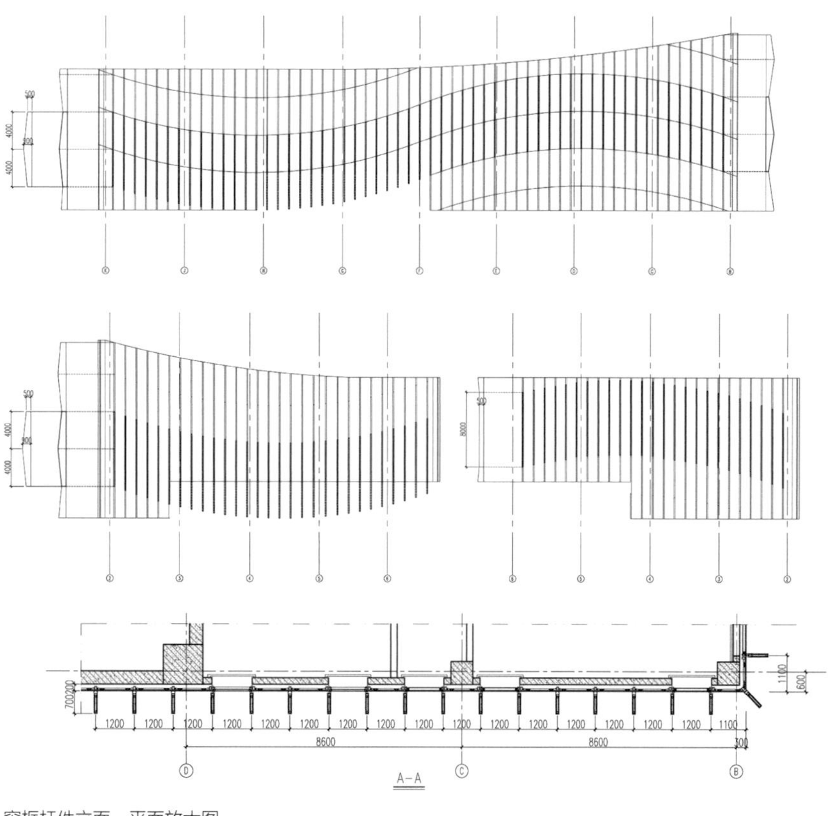

窗框杆件立面、平面放大图
Window Frame Bar Elevation, Plane Enlargement

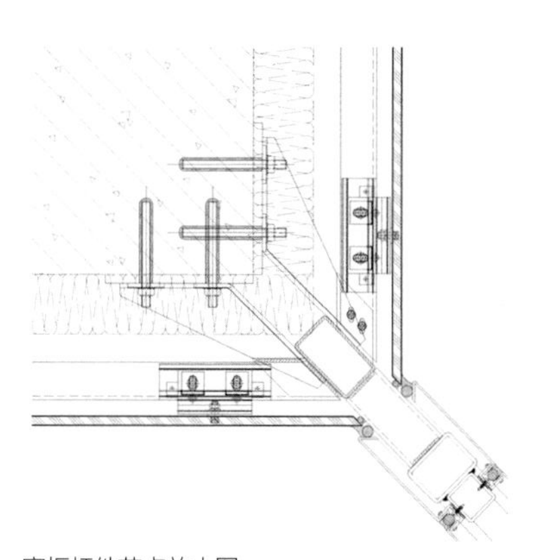

窗框杆件节点放大图
Window Frame Bar Node Enlarged View

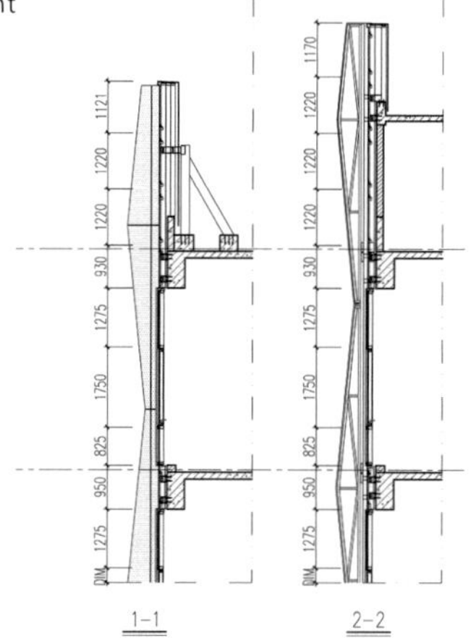

窗框杆件剖面放大图
Window Frame Bar Section Enlarged View

好用的产品
User-friendly Products

重新树立的设计价值观 / 上海绿地创新产业中心
Re-established Design Values / Greenland Innovative Industry Center of Shanghai

专业与非专业 / 上海海博西郊冷链物流园
Professional and Unprofessional / Shanghai Haibo Western Suburbs Cold Chain Logistics Park

适应性设计 / 上海南翔小绵羊总部产业园
Adaptive Design / Industrial Park of Nanxiang Small Sheep Headquarter in Shanghai

外科手术式改造 / 上海蓝天数字游戏产业园
Surgical Transformation / Blue Sky Digital Game Industry Park in Shanghai

《设计心理学》的作者唐纳德·诺曼有句名言——在人和设计之间，人是不会错的，错的只有设计。即使最新一版也时隔 20 年，但其中关注用户的核心理念至今仍十分受用。

以往人们都觉得设计师艺术，是创意，而在该书中，诺曼的定义则不同。他认为设计的本质是沟通，是一门设计者和使用者之间，通过产品实现无声沟通的学问。汽车门上的车窗按钮、卫生纸上的虚线，易拉罐上的拉环、iPhone 的 home 键，无一不是既简洁美观又上手易用的设计，完全不用说明书教。就如作者所言，面对一个产品时，弄不清楚的一定是设计的问题。

为什么一进酒店客房，我们发现马桶边的卫生纸都被折成了一个三角形的头？就是为了在没有服务员解释的情况下，让客人知道这间客房是打扫整理过的。为什么医院的环境大都是白色的？就是为了让人一眼就能理解到这里是清洁卫生的场所。

仅有创意构不成设计，因为设计的本质是建立创意与用户的联系，而好用的产品则是这段桥梁。

2

Galerija
Gallery
ギャラリー

耳桥 [以耳为桥 · 聆取天音]
"Ear Bridge " The Ear is the Bridge, Listen to Voice

张佾媛
Yiyuan，Zhang

镜片，麻纸，水墨、碳
Lenses，Hemp Paper, Ink, Carbon

原尺寸 Original Size 33mm × 33mm

2018

重新树立的设计价值观 / 上海绿地创新产业中心
Re-established Design Values / Greenland Innovative Industry Center of Shanghai

上海

占地面积	：27 966 平方米
建筑面积	：62 200 平方米
建筑高度	：24 米
设计时间	：2014

Shanghai

Floor Space	: 27,966 m^2
Gross Floor Area	: 62,200 m^2
Building Height	: 24 m
The Time of Design	: 2014

绿地花园办公实践
The Practice of Work of Green Garden

十年前，我工作的地方是一个有院子的小独栋，在工作的同时，仍保持与自然的互动，观察植物的变化并感知四季。而今上海高强度高密度的城市发展下，这样的场景几乎是一种奢望。所以，当遇到这块定位为“花园办公”的土地时，我们希望建立植物与人的美好互动，人人都能在花园中办公……带着这样的愿景，我们开始了设计。

这是绿地在 2014 年的一个项目，当时正准备赶超万科登顶房地产榜首，节奏很赶，要求快速落地，要求产品创新有力度，并要面对凯达团队的竞争。如何以缜密的逻辑推演掌控设计节奏，如何面对绿地这样经验丰富的地产开发商提出产品和价值的创新，如何将有价值有情怀的空间故事化地呈现市场运营的策略，是我们探讨及学习的重点。

该项目位于上海市宝山国际研发总部商务区，中环与外环之间，临沪太路，上大路地铁上盖，周围分布大学、科研机构，人才集聚，具备大学、社区、产业园联动发展的潜力，同时，基地毗邻河流和城市绿地，景观资源优秀。项目周边产业园区集群已逐步形成，但由于临近宝山、普陀、老闸北交汇处，环境品质的印象并不高。所以项目品质的打造至关重要，而产品如何在成熟的市场圈中脱颖而出也是考验点。

Ten years ago, the place where I worked was a small independent building with a courtyard. While working, I kept interacting with nature, observing changes of plants to perceive the four seasons. With the development of the city of high-intensity and high-density in Shanghai today, such a scenario is almost a luxury. Therefore, when we encounter this land that is defined as a "garden office", we hope to establish a beautiful interaction between plants and people. Everyone can work in the garden... With the hope, we started the design.

This was a project of Greenland in 2014. At that time, it was preparing to surpass Vanke to go the top the market of real estate. The pace was very fast. It required quick landing and powerful product innovation, facing the competition from the Al-Qaeda team. How to control the rhythm of design with careful logic, how to face the developers of green real estate, putting forward the innovation of product and value, how to let the space that has value and feelings present the strategy of market operations by stories? They are the key we should discuss and learn.

The project is located in Baoshan International R & D headquarters business district in Shanghai, between Central and Outer Ring, near Hutai Road, and covered by Shangda Road subway. It is surrounded by universities, scientific research institutions, and talented people. It has a potential of joint development of universities, communities and industrial parks. At the same time, the base is adjacent to rivers and urban green areas. The resources of landscape are excellent. The clustering of industrial parks around the project has gradually taken shape. However, because it is near to the intersection of Baoshan, Putuo and Laozhabei, the impression of environmental quality is not high. Therefore, the construction of project quality is crucial. How the product stands out in the mature market circle is also a test point.

中环、地铁上盖、临主干道、一线开发商
高密度、低容积率、强限高
定制创新花园办公产品
2017 年地产设计大奖 · 优秀奖
第 12 届金盘奖最佳产业地产奖

Central、Subway cover、Pro Main Road、Front-line developers
High-density、Low volume rate、Strong Limit High
Customized Innovation Garden Office products
2017 Real Estate Design China Award · Excellence Award
The 12th Golden Plate Award for Best Industrial Property Award

方案竞标单位：凯达环球
Project Bidding unit: Aedas

景观设计：AECOM 、纳千
Landscape Design: AECOM、LGWS DESIGN,LLC

室内设计：集艾室内设计（上海）有限公司
Interior Design: G&A Interior Design

技术深化：上海中森建筑与工程设计顾问有限公司
Technology Deepening:Johnson-cadg Architecture&Engineering Design

幕墙设计：上海安雷幕墙工程（顾问）有限公司
Curtain Wall Design:Umbreal Curtain Wall Engineering

灯光设计：上海麦索照明设计咨询有限公司
Lighting Design: Meico Lighting Design Consulting

31°11′
25.83″ N
121°25′
8.61″ E

The Main Core & The Design Method Is Value Creation

Land area 27,966m²
Density 45% FAR 1.5

由上而下的规划价值

最好的资源留给最好的产品，是地产开发中资源配置的铁则。而规划的作用在于平衡资源均好性和公共空间利益。将资源渗透入地块，组织合理的资源配置网络，是本案规划的切入点。

该项目最重要的资源就是东侧的公共绿地和西侧的地铁出入口。设计上首先将公共绿地与地铁出入口直接串联，将视线通廊由西向东延伸，贯穿整个地块，形成开放的景观主轴。同时，西侧沪太路虽是重要的入口形象面，上位规划禁止设置机动车出入口，为了平衡入口形象和交通效率，规划将景观打造融于其中。步行出入口设在东西两侧，作为景观主轴的端口。车行出入口设在南北两侧，连接形成景观次轴。最终以黄金十字的布局，将地块分为相对独立的四大组团，奠定价值分配所依托的规划格局。

其次通过对地块自身及周边景观资源划分等级，产生由高到低三种等级景观区：东侧公共绿地为 A 级景观区，景观轴交汇的中心广场为 B 级景观区，景观轴分割出的四大组团为 C 级景观区。遵循资源配置的原则，我们将产品分为 L/M/S 三档，对应布置在三种等级的景观区内：A 级景观区配置 L 型 2 000 平方米 产品，B 级景观区配置 M 型 1 500 平方米产品，C 级景观区配置 S 型 1 000 平方米产品，最终实现由资源引导的规划布局。

The Value of Planning from Top to Bottom

Best resources are reserved for best products, which is the iron standard for resource allocation in development of real estate. The role of planning is to balance goodness of resources and benefits of public space. The penetration of resources into land plots and the organization of a reasonable resource to deploy network are the starting point.
The most important resources are the public green space on the east and the entrance and exit of subway on the west. The public greenbelt will be directly connected with the entrance and exit. Line of sight corridor will extend from west to east, running through the entire plot, forming an open main axis of landscape. Meanwhile, although Hutai Road on the west is an important image of entrance, the upper plan prohibits the setting of entrances and exits of motor vehicles. To balance the image of entrance and efficiency of traffic, the plan will integrate the landscape into it. Walking entrances and exits, as the port for the main axis of landscape, are located on both sides of the east and west. Vehicle entrances are located on both sides of the north and south, connecting to form the secondary axis of landscape. Finally, with the layout of Golden Cross, the land will be divided into four groups that are relatively independent, which lays the foundation of the planing layout for distribution of values.
Secondly, by classifying the land itself and its surrounding resources of landscape, areas of landscape of three levels from high to low are produced: the eastern public green space is a landscape area of Grade A, and the central square where the landscape axes meet is a landscape area of Grade B. The four groups that are landscape areas of Grade C is segmented out by the landscape axis. Following the principle of resource allocation, we divided products into L/M/S third gear, which is arranged in the areas of landscape of three levels: A-level landscape area is deployed with L-type product of 2000 square meters. B-level is deployed with M-type product with 1500 square meters. C-level is equipped with S-type product of 1,000 square meters. The resource-guided layout of programme will be realized finally.

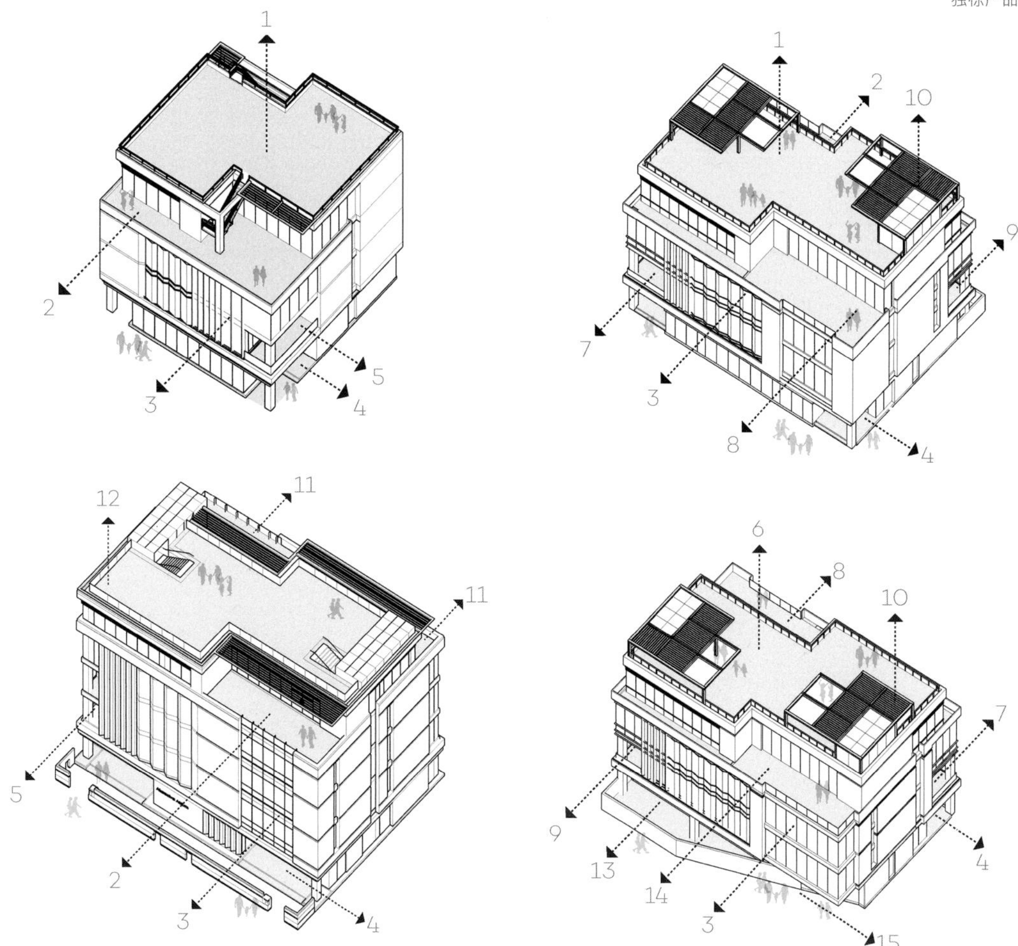

1 屋顶绿化
丰富办公生活
Roof Greening
Enrich Office Life

2 大面积露台
正对主轴或庭院
体现价值导向
Large Area Terrace
On the Spindle or the Courtyard
Reflect the Value Orientation

3 大面积玻璃
具有价值导向的立面设计
Large Area Glass
Value - Oriented Facade Design

4 灰空间
引入庭院绿色
Grey Space Introducing
Courtyard Greening

5 阳台
享受庭院绿化
Balcony,Enjoy
the Courtyard Greening

6 四层屋顶绿化
最大化享受景观主轴
丰富办公空间
Four-storey Roof Greening
Maximizing the Enjoyment
of City Parks Rich Office Space

7 阳台
享受庭院绿化
Balcony Enjoy
Courtyard Greening

8 大面积露台
享受庭院绿化
Large Area Terrace
Enjoy Courtyard Greening

9 阳台
承受主轴景观
Balcony Bearing the
Main Axis Landscape

10 出顶楼梯、构架
提升景观享受品质
Top Stair, Frame Enhance the
Quality of Landscape Enjoyment

11 大面积露台
正对城市公园
体现价值导向
Large Area Terrace
Right on the City Park
Embody Value Orientation

12 五层屋顶绿化
最大化享受城市公园
丰富办公空间
Five-storey Roof Greening
Maximizing the Enjoyment of
City Parks Rich Office Space

13 私有庭院
享受户外办公与咖啡
Private Courtyard Enjoy
Outdoor Office and Coffee

14 大面积露台
正对景观主轴
体现价值导向
Large Area Terrace
Facing the Landscape Spindle
Reflect the Value Orientation

15 灰空间
享受室外办公
Gray Spaces
Enjoy Outdoor Work

视线分析 Vision Analysis

垂直向通透的视线
Perpendicular to the Transparent Sight

方向受控制的视线
Direction of the Controlled Line of Sight

完全通透的开放面
Fully Permeable Open Surface

视线控制的开放面
Open Side of Vision Control

附带窗洞的石墙面
Stone Wall with Window

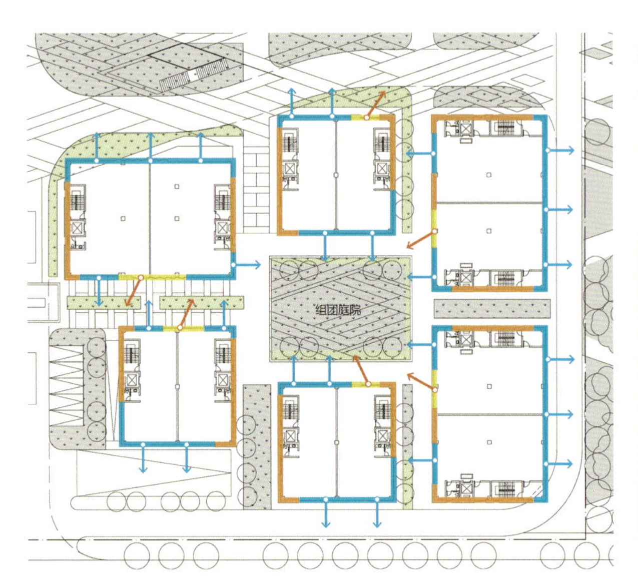

立面也是产品

设计不仅在平面上对应资源分级，而且在立面设计中也遵循价值逻辑的导向。朝向大尺度公共景观，建筑打开通透界面，而向较小尺度的中庭则引导性开窗，通过竖向带有角度的百叶等半通透界面引导视线。核心筒布置在建筑间距较小的山墙面，既避免视线干扰，又可以让视线通透的空间更加完整方正，为办公单元的可分可合留有余地。立面通过将平面上的资源空间布局立体化，作为产品，提供价值的另一道驱动力。

另外，价值不仅体现在空间的品质上，还体现在时间的联系上。作为与企业共同生长的园区，空间扩张与企业升级的同步是必须解决的问题。我们以生长为理念，提供了一家公司在园区中十年左右的生长空间。从最初联合办公的几张办公桌到一间办公室，到租赁一间 S 型独栋，再到转移至更大型的独栋，再到占据半个甚至整个组团，这间园区将为企业定制终身升级服务，朝夕相伴共同成长。

Facade is also a Product

Design not only corresponds to resource grading in the plane, but also follows the guidance of logic of value in the facade design. Facing the large-scale public landscape, the building opens up the transparent interface and guide the window open to the smaller-scale atrium, guiding the line of sight through semi-transparent interfaces such as vertical angled louvers. The core and easy arrangement is on a gable wall with a small distance between buildings, which not only avoids the disturbance of sightlines, but also makes the space that is transparent to sight more complete and square, providing room for the separability of office units. Facades, as products, provide another driving force for value by making a three-dimensional layout of resources on the plane.

In addition, value is not only reflected in the quality of space, but also in the connection of time. As a park that grows with enterprises, the synchronization between expansion of space and upgrade of enterprises is a problem that must be solved. Our concept of growth provides a space for a company to grow in the park for a decade or so. From a few office desks that are co-organized initially to an office, to the lease of a S-type detached house, to a larger detached house, to occupying half or even the entire group, the park will customize lifelong services of upgrading for enterprises, accompanying with each other from morning to night and growing together.

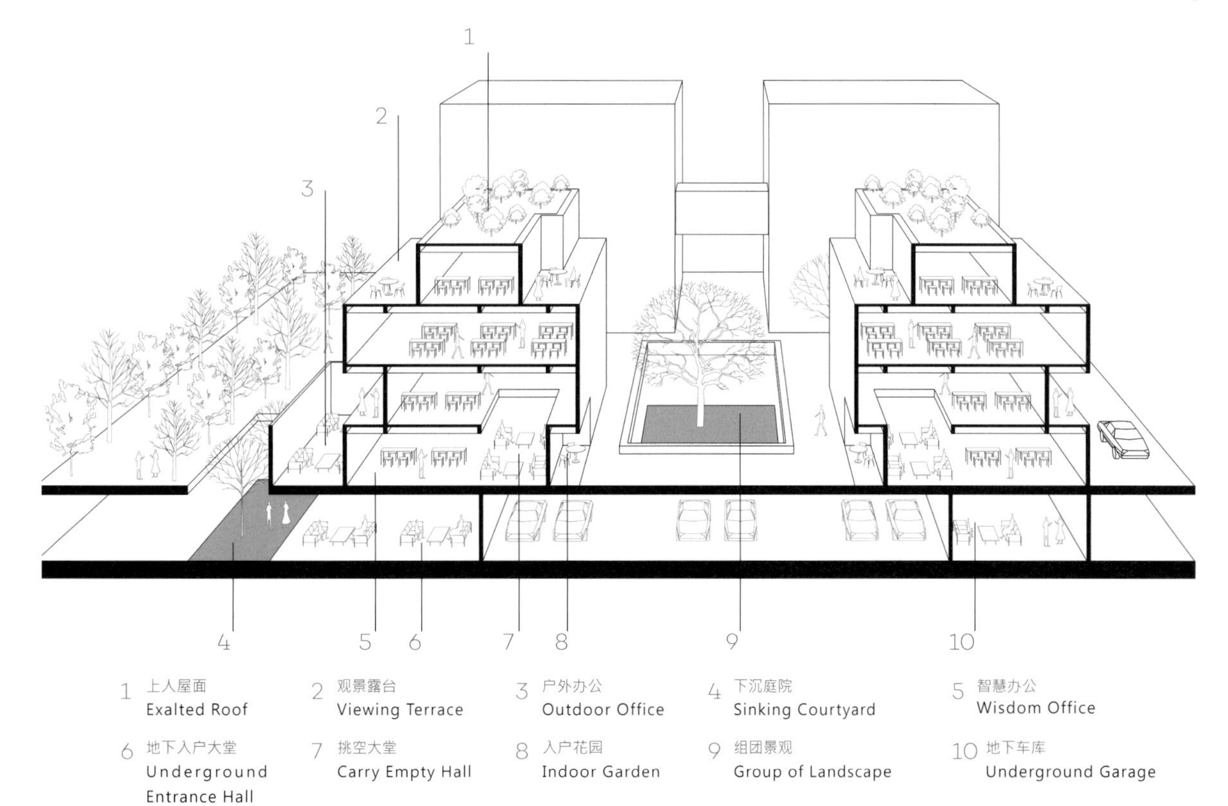

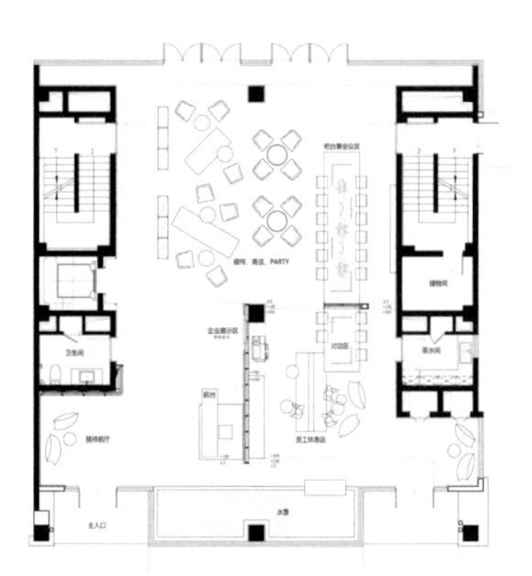

1 000 ㎡小户型独栋
一层平面

1,000 ㎡ Small Apartment Single
a Layer of Plane

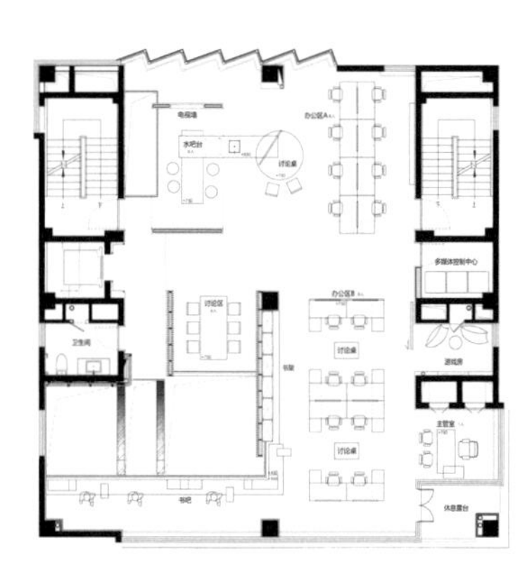

1 000 ㎡小户型独栋
二层平面

1,000 ㎡ Small Apartment Single
Two-Level Plane

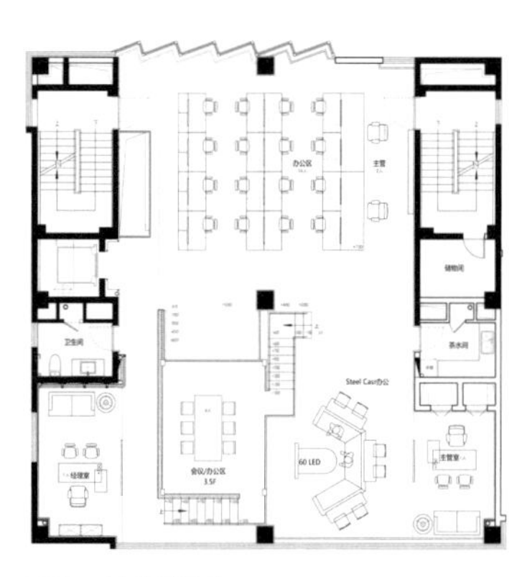

1 000 ㎡小户型独栋
三层平面

1,000 ㎡ Small Apartment Single
Three-Level Plane

建造过程，摄于 2017 年 3 月
Construction Process,2017.3

立面造型细部
Facade Modeling Detail

室内样板效果
Interior Model Effect

室内样板房，摄于 2016 年 11 月
Interior Model Room,2016.11

开发节奏与建造控制

地产类开发往往是快速且富有挑战的：即便在后期施工图阶段，由于成本及定位的多次调整，材料经历石材——体板—— 涂料——体板——石材等多轮反复，并要求设计快速回应，甚至当晚开会有意见，次日上午就要求方案调整出成果，我们也遇到这种情况。即便如此，实施过程中尽可能的做到精细控制：材料选型、上墙看样、分缝与交接设计经过反复推敲；格栅百叶的密度与角度通过实体模型的制作确定比例；深度制作立面控制手册；参与景观、室内、灯光专业设计等。

好的设计资源都是有限的，而建筑师的价值正在于为其赋能；
产品只有在产品线上才有市场活力，又会有真正的价值升级；
价值的核心在于资源与品质，价值创造的核心在于灵感与逻辑。

The Development of Rhythm and Controlling of Construction

The development of real estate is often fast and challenging: even at the later stage of construction drawings, due to multiple adjustments in costs and positioning, the material undergoes many repeated rounds, stone-body board-paint-body board-stone, requiring rapid response of design. Indeed, if there were opinions in the evening meeting, results that are adjusted by plans are required the next morning. We also encountered this situation. Even so, fine control was implemented in the process as far as possible: material selection, sample look on the wall, splitting and handing over design were repeatedly scrutinized; the proportion of the density and angle of the grille determined by the production of the solid model; making the manual of facade control deeply; participating in the professional design of landscape, interior, lighting and so on.

Good resources of design are limited. And the value of architects is empowering them. Products only have market vitality in the line of product and there will be real escalation of value; the core of value lies in resources and quality and the core of creation of value is inspiration and logic.

沿上大路照片，建造过程，摄于 2016 年 6 月
Photo By Shangda Road , The Construction Process,2016.07

开工照片 摄于 2017,11
The Start of the Photos is Taken In 2017.11

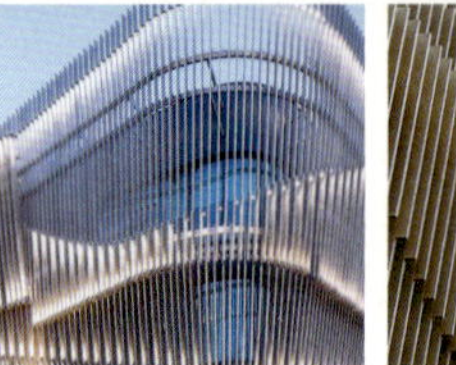

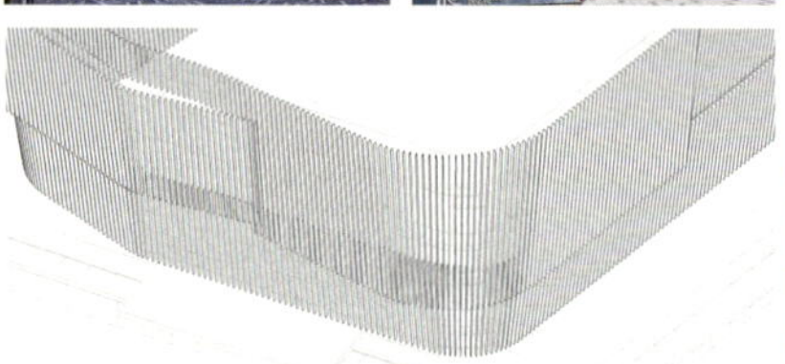

与工期同步推进的模型推敲
Simultaneous Advancement of Model Assessment

内部空间关系
户间转换灵活性
Internal Spatial Relation
Inter-household Conversion
Flexibility

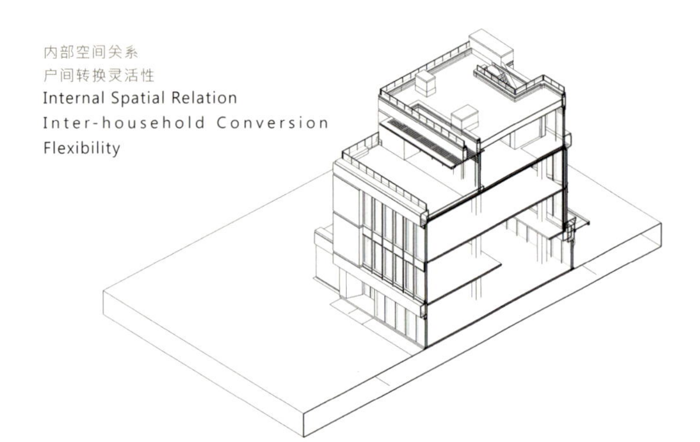

组团间连廊
户间生长延展性
Intergroup Corridor
Growth Ductility Among
Households

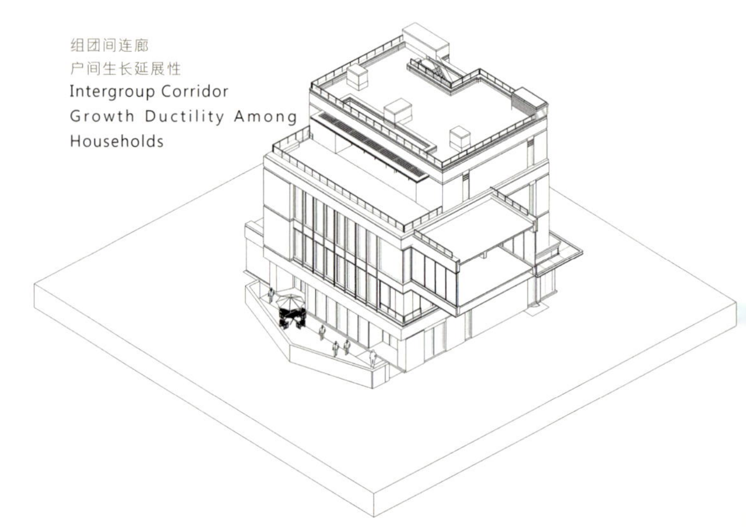

脚手架拆除后场景
The Scene after the Scaffold is Dismantled

以生长为理念

每家公司规模发展中，人员扩张与空间升级是必须解决的问题，设计上通过，提供了一家公司十年左右的生长空间。从最初联合办公的几间甚至一间办公室，到租赁一间 S 型独栋，再到更大独栋，再到半个甚至整个组团，均为其提供各种办公空间组合模式。

密度 45%、容积率 1.5、限高 24m 的紧凑地块中，结合绿化的宜人空间是品质的最佳体现。为了营造可以自由舒展的园区，我们将景观渗透，打造花园办公产品。在花园办公中，每个院落都有自己的独立入口，每家办公都能拥抱院子与阳光。环廊合抱，露台独享，在风声虫鸣中找回思考的安静，才是东方花园的魅力。而在大场，我们结合空间起伏收放，营造出下沉庭院、挑空大堂、室外会议区、檐下门厅等一系列高溢价场所，将这份文化意趣与空间产品结合，创造更高的价值。

Regarding Growing as a Concept

In the development of the scale of each company, personnel expansion and upgrading of space are problems that must be solved. The design provides a space for a company to grow in the park for a decade or so. From a few office desks that are co-organized initially to an office, to the lease of a S-type detached house, to a larger detached house, to occupying half or even the entire group, all provide a variety of modes of combination of office space.

In a compact plot with a density of 45%, a floor area ratio of 1.5 and a limited height of 24 meters, the pleasant space combined with greening is the best expression of quality. In order to create a park that can be freely stretched, we infiltrate the landscape, creating products of garden office. In the garden office, each courtyard has its own independent entrance, and every office can embrace the yard and the sun. It is the charm of the oriental garden that it is held together in a ring of galleries, enjoying the platform solely and finding the quietness of thinking in the sound of wind and birds. In the big field, we combine fluctuation of space to create a series of high premium places, such as sinking courtyards, empty lobby, outdoor meeting area, and underneath hall. We combine this cultural interest with products of space to create higher-level value.

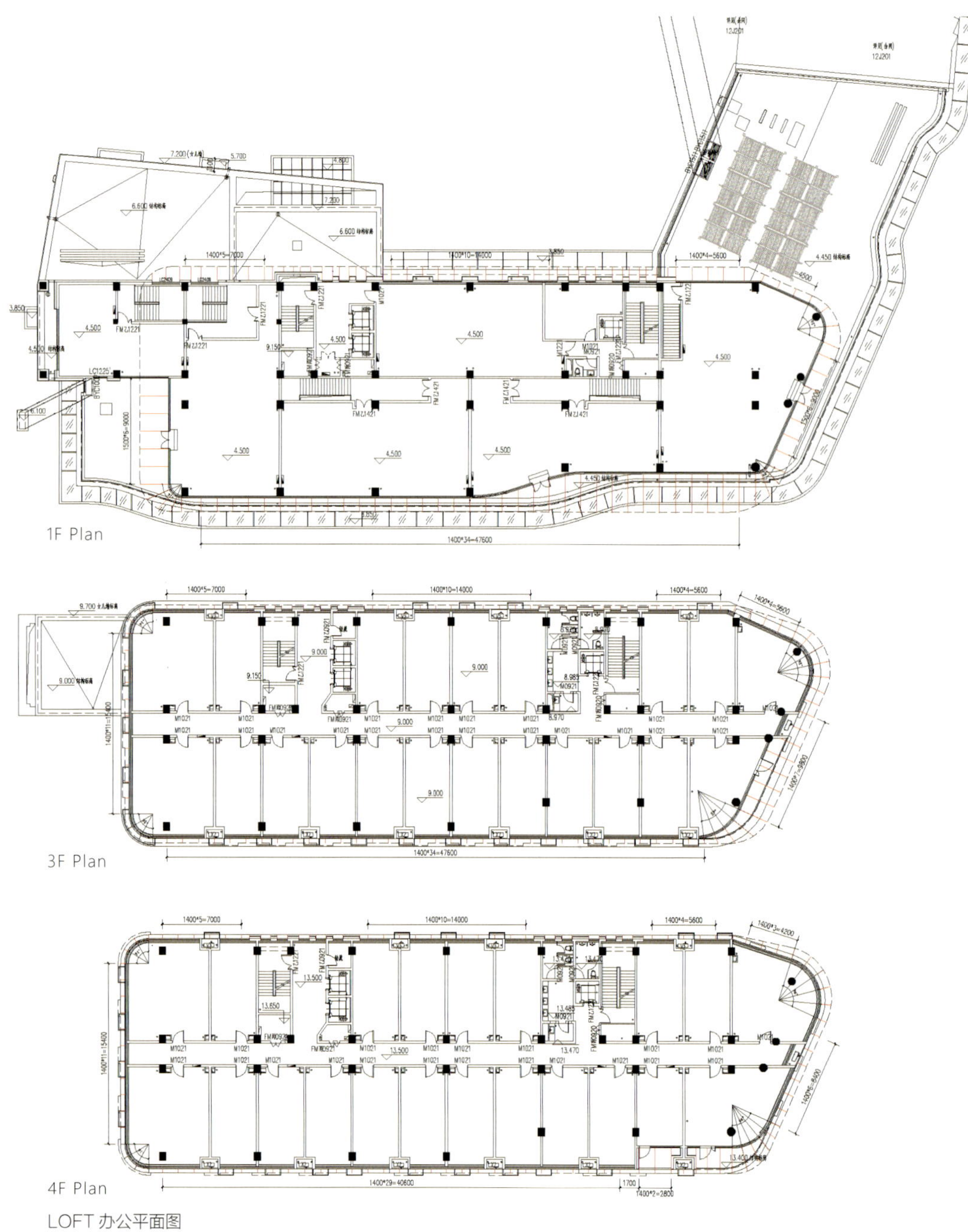

LOFT 办公平面图
Loft Office Plan

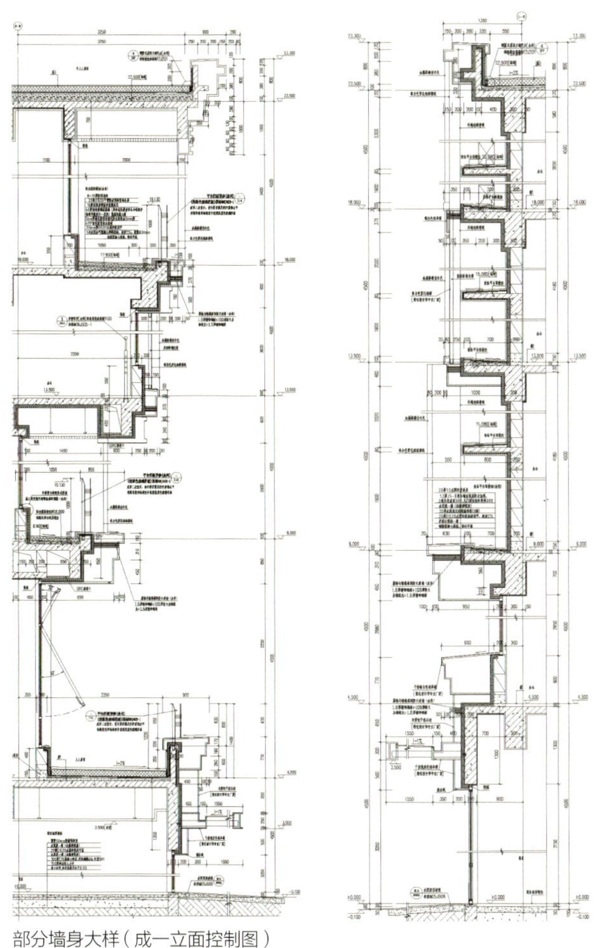

部分墙身大样（成一立面控制图）
Part Of The Wall Detail (Elevation Control Chart by One Studio)

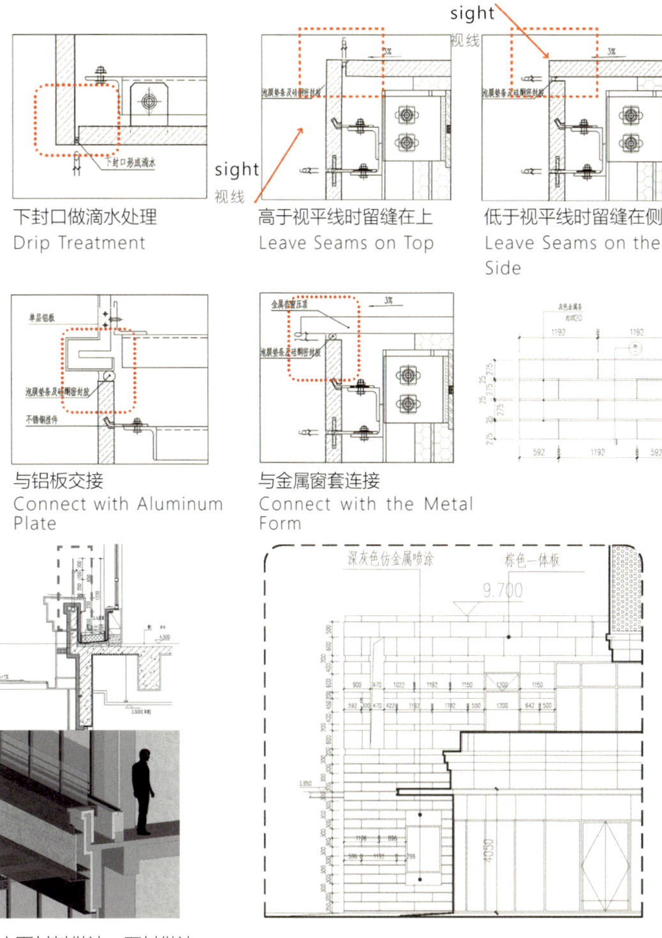

下封口做滴水处理
Drip Treatment

高于视平线时留缝在上
Leave Seams on Top

低于视平线时留缝在侧
Leave Seams on the Side

与铝板交接
Connect with Aluminum Plate

与金属窗套连接
Connect with the Metal Form

立面材料做法—石材做法
The Approach of Facade Material—Stone

分缝原则

主体以单边 1 400mm 为标准段（取窗框中心线）
弧线端从圆心发散等角度分缝
核心筒外墙处铝板分缝与墙体保持分缝一致
局部特殊段以短边不超过 1 500mm 控制尺寸
吊顶分缝投影线为二层阳台吊顶分缝和层间线脚分缝。
缝宽 6mm，与相接层间线脚分缝一致（余同）

The Principle of Parting

The main body is the single side 1,400mm as the standard section (take the center line of the window frame)
Arc end parting from the center angle of divergence
The core tube at the external wall and the wall to maintain consistent parting parting plate
Local special section with short edge no more than 1,500 control dimensions
The parting line is two ceiling projection balcony ceiling parting layer and molding part.
Width of 6mm are connected with the interlayer joint agreement (all the same).

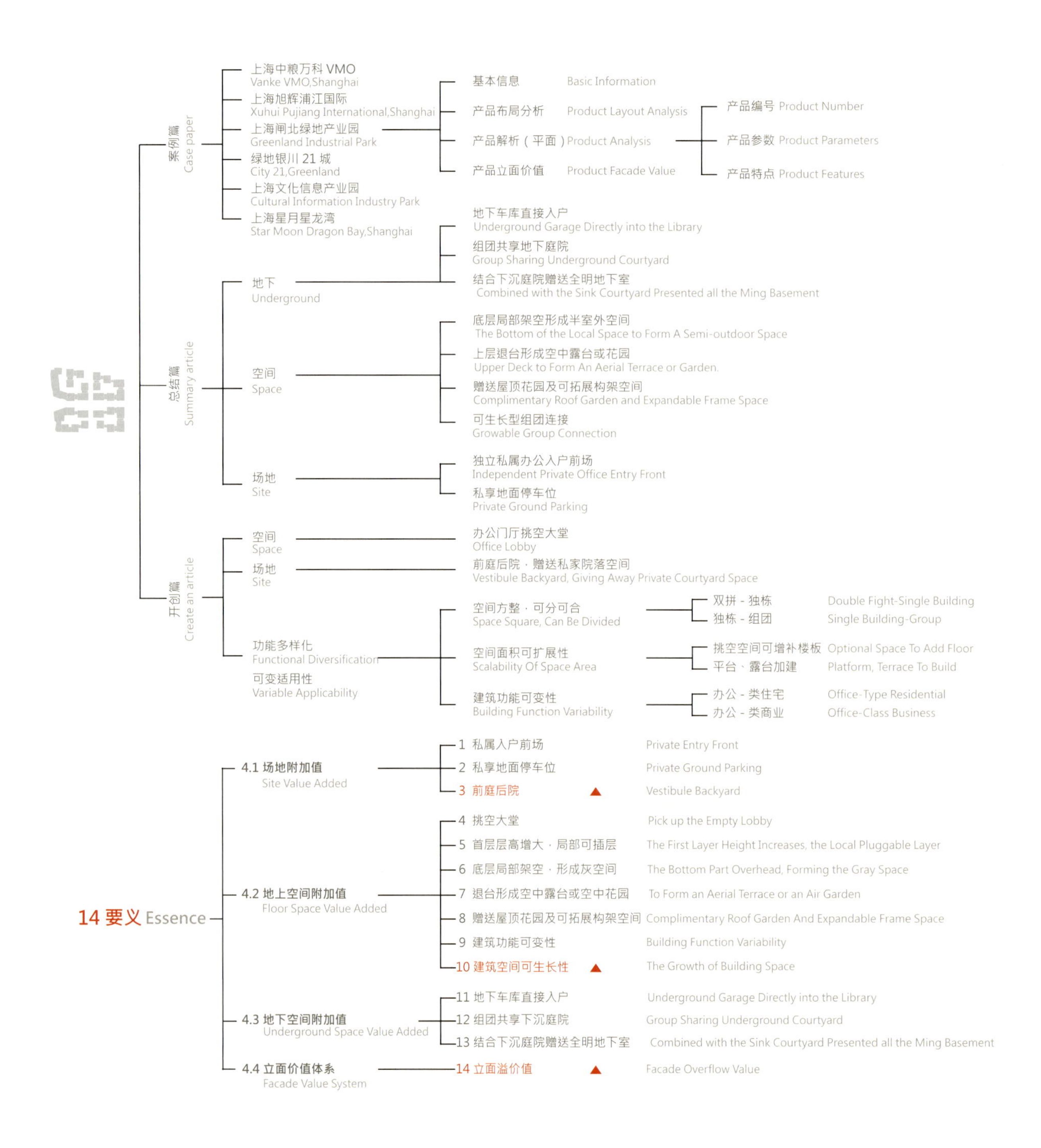
案例篇 Case paper
上海中粮万科 VMO
Vanke VMO,Shanghai
上海旭辉浦江国际
Xuhui Pujiang International,Shanghai
上海闸北绿地产业园
Greenland Industrial Park
绿地银川 21 城
City 21,Greenland
上海文化信息产业园
Cultural Information Industry Park
上海星月星龙湾
Star Moon Dragon Bay,Shanghai
基本信息 Basic Information
产品布局分析 Product Layout Analysis
产品解析（平面）Product Analysis
产品立面价值 Product Facade Value
产品编号 Product Number
产品参数 Product Parameters
产品特点 Product Features
总结篇 Summary article
地下
Underground
地下车库直接入户
Underground Garage Directly into the Library
组团共享地下庭院
Group Sharing Underground Courtyard
结合下沉庭院赠送全明地下室
Combined with the Sink Courtyard Presented all the Ming Basement
空间
Space
底层局部架空形成半室外空间
The Bottom of the Local Space to Form A Semi-outdoor Space
上层退台形成空中露台或花园
Upper Deck to Form An Aerial Terrace or Garden.
赠送屋顶花园及可拓展构架空间
Complimentary Roof Garden and Expandable Frame Space
可生长型组团连接
Growable Group Connection
场地
Site
独立私属办公入户前场
Independent Private Office Entry Front
私享地面停车位
Private Ground Parking
开创篇 Create an article
空间
Space
办公门厅挑空大堂
Office Lobby
场地
Site
前庭后院 · 赠送私家院落空间
Vestibule Backyard, Giving Away Private Courtyard Space
功能多样化
Functional Diversification
可变适用性
Variable Applicability
空间方整 · 可分可合
Space Square, Can Be Divided
双拼 - 独栋 Double Fight-Single Building
独栋 - 组团 Single Building-Group
空间面积可扩展性
Scalability Of Space Area
挑空空间可增补楼板 Optional Space To Add Floor
平台 、露台加建 Platform, Terrace To Build
建筑功能可变性
Building Function Variability
办公 - 类住宅 Office-Type Residential
办公 - 类商业 Office-Class Business
14 要义 Essence
4.1 场地附加值
Site Value Added
1 私属入户前场 Private Entry Front
2 私享地面停车位 Private Ground Parking
3 前庭后院 Vestibule Backyard
4.2 地上空间附加值
Floor Space Value Added
4 挑空大堂 Pick up the Empty Lobby
5 首层层高增大 · 局部可插层 The First Layer Height Increases, the Local Pluggable Layer
6 底层局部架空 · 形成灰空间 The Bottom Part Overhead, Forming the Gray Space
7 退台形成空中露台或空中花园 To Form an Aerial Terrace or an Air Garden
8 赠送屋顶花园及可拓展构架空间 Complimentary Roof Garden And Expandable Frame Space
9 建筑功能可变性 Building Function Variability
10 建筑空间可生长性 The Growth of Building Space
4.3 地下空间附加值
Underground Space Value Added
11 地下车库直接入户 Underground Garage Directly into the Library
12 组团共享下沉庭院 Group Sharing Underground Courtyard
13 结合下沉庭院赠送全明地下室 Combined with the Sink Courtyard Presented all the Ming Basement
4.4 立面价值体系
Facade Value System
14 立面溢价值 Facade Overflow Value

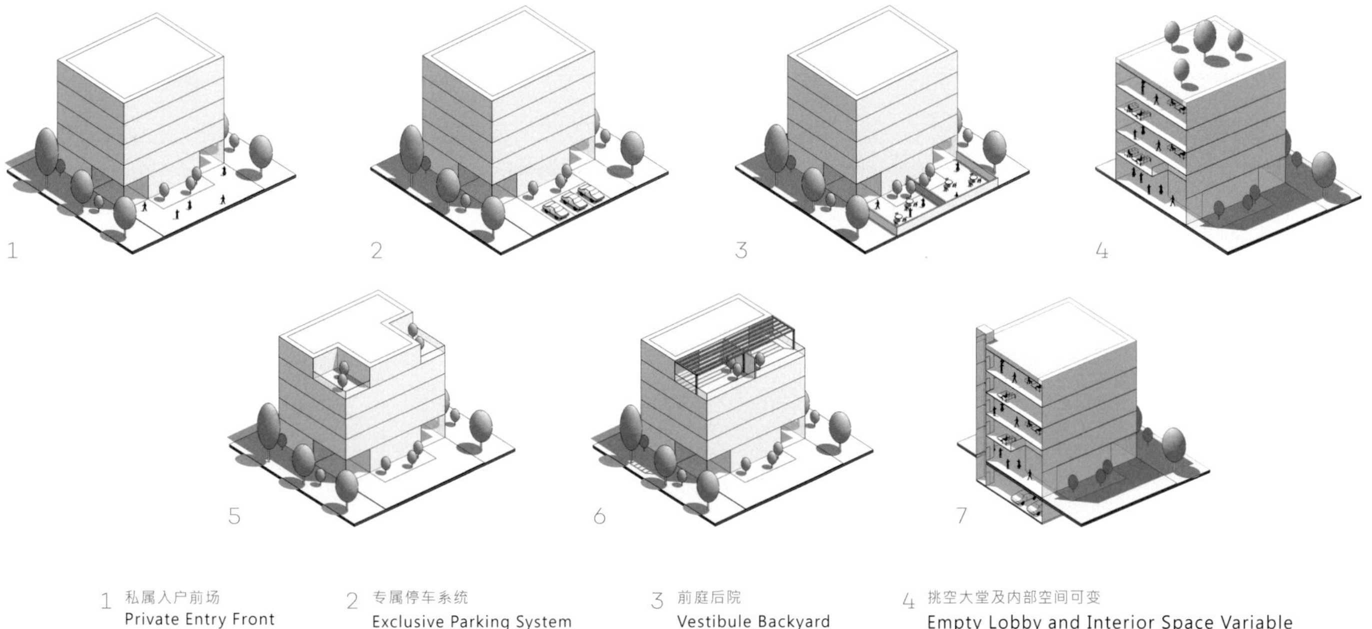

1 私属入户前场
Private Entry Front

2 专属停车系统
Exclusive Parking System

3 前庭后院
Vestibule Backyard

4 挑空大堂及内部空间可变
Empty Lobby and Interior Space Variable

5 退台形成空中露台或空中花园
To Form an Aerial Terrace or an Air Garden

6 赠送屋顶花园及可拓展构架空间
Complimentary Roof Garden and Expandable Frame Space

7 地下车库直接入户
Underground Garage Directly into the Library

产品线研究与研发

业主要求该项目在市场上既有独一无二的特质，又有可更迭推广的模板。所以我们从市场调研出发，归纳设计成果，形成了花园式办公产业园区设计的十三要义，积累了此类项目的操作经验，其推广在市场上也形成了较大的影响力。

常规办公产品将核心筒中置，强调双拼产品的独立性。而我们在设计中，从企业发展的生长性和灵活性入手，偏置核心筒，从而中间随时可打通“扩容”。受硅谷 “户外办公”和法租界洋房花园的启发，在上海这种户外环境相对宜人的地域，我们结合灰空间设计了多花园互动体验场所。其他诸如资源分级、露台阳台、屋顶花园、地下利用等比较成熟的设计手法，也作为提高溢价的手段。

最终我们在快节奏中完成了一次强营销、高价值、全周期的设计。即使我们当时作为一个缺乏经验的年轻团队挑战重重，但我们在完成业主诉求的同时，创造了花园办公的典范，荣获金盘奖 2017 年上海地区最佳园区。

Research and Development of Product Line

The project is required by proprietors to have unique characteristics in the market and templates that can be promoted. Therefore, starting from the market research, we sum up the results of design, forming the thirteen essential principles of the design of the garden-style office industry park. We have accumulated the experience of operating of such projects whose promotion has also formed a large influence on the market.

The core of conventional office products will be placed in a nutshell, emphasizing the independence of the double products. In the design, we begin with the growth and flexibility of development of enterprises and offset the core simplicity so that we can open up the "expansion" at any time in the middle. Inspired by the"outdoor office"of Silicon Valley and foreign-style houses in French concession, in the relatively pleasant outdoor environment in Shanghai, we designed a multi-garden interactive experience venue in combination with gray space. Other mature techniques of design , such as resource grading, typhoon balcony, roof garden, and underground use, are also serve as means to increase premiums.

In the end, we completed a strong marketing, high-value and full-cycle design in a fast pace. Although We, as an inexperienced young team, have much challenge at that time, we created a model for the garden office and won the Jinpan Award for the best park in Shanghai in 2017 when we finished appeals of proprietors.

专业与非专业 / 上海海博西郊冷链物流园
Professional and Unprofessional / Haibo Western Suburbs Cold Chain Logistics Park in Shanghai

上海

占地面积 ： 45 873 平方米

建筑面积 ： 58 000 平方米

建筑高度 ： 20.2 米

设计时间 ： 2017

专业能力就是能将复杂问题规范化高效处理的能力，而专业素养就是能在动态中掌握体系核心的素养。陌生并不可怕，可怕的是生疏，对于专业人员来说，即使面对非专业的领域，同样可以通过类比、学习和沟通打破壁垒，海博冷链物流园就是我们一次以专业的思维尝试非专业类型的设计。

这是光明集团的一个战略性项目。以“冷链物流”为核心和特色功能，“冷链”和“物流”关联产业为基本载体，其他生产性服务业为辅助配套，现代信息和物流网为支撑，产业 + 商务为驱动引擎，光明希望打造创新示范性的“冷链物流产业集聚区”。

项目主要功能由物流区和商办配套区构成。物流区包含高层低温库、单层低温库、多层常温库、单层常温库、后勤用房、制冷机房等单体建筑，建筑功能性要求极高。项目位于西虹桥待开发地段，受周边道路宽度限制，交通高效化运作也被列为设计的难题。在一期设计中造型的处理欠缺考虑，在成本可控的条件下，综合提升巨构建筑形象也成为重要课题。

在深入剖析项目后，我们从模式切入，配合策划及相关专业，吸取先进案例成功要素，并代入项目开发中成熟的商业思维，在研习摸索中催生创意。

Shanghai

Floor Space : 45,873 m^2

Gross Floor Area : 58,000 m^2

Building Height : 20.2 m

The Time of Design : 2017

Professional competence is the ability to handle complicated problems in a standardized and efficient manner. Shitsuke of profession is the ability to grasp the core qualities of the system in a dynamic manner. The strangeness is not terrible while disacquaintance is fearful. For professionals, even facing non-professional fields, they can also break barriers through analogy, learning and communication. Haibo Park Logistics of Cold Chain a design that is a non-professional type that we try by professional thinking.

This is a strategic project of Bright Group. "Cold-chain logistics" is the core and characteristic function. "Cold-chain" and "logistics"-related industries are the basic carriers. Other industries of production service are supplementary facilities. Modern information and logistics networks are supports. Industry + Business is the driving engine. Bright hopes to create innovative and exemplary "industrial clusters of cold-chain logistics ".

The main functions of the project consist of logistics areas and supporting areas for business and office. The logistics area includes high-rise storage of low-temperature, single-layer storage of low-temperature, multi-layer storage of normal temperature, single-layer storage of normal-temperature, rooms of rear services, and room of refrigerating machine room. The building has high requirements of function. The project is located in the area of West Hong Bridge that is to be developed. Limited by the width of surrounding roads, the operation of efficient traffic is also listed as a problem of design. In the first phase of the design, the handling of the modelling is lack of consideration. Under the condition of controllable costs, it has become an important issue to comprehensively upgrade the image of the giant building.

After an deep analysis of the project, we draw on the successful elements of advanced cases, learning from the model, coordinated with the planning and related majors, and integrate the mature ideas of business into the development project, promoting creativity in the process of study and grope.

非传统行业型模式创新

The Model Innovation of Non-traditional Industry

31°13′
13.3″ N
121°14′
44.6″ E

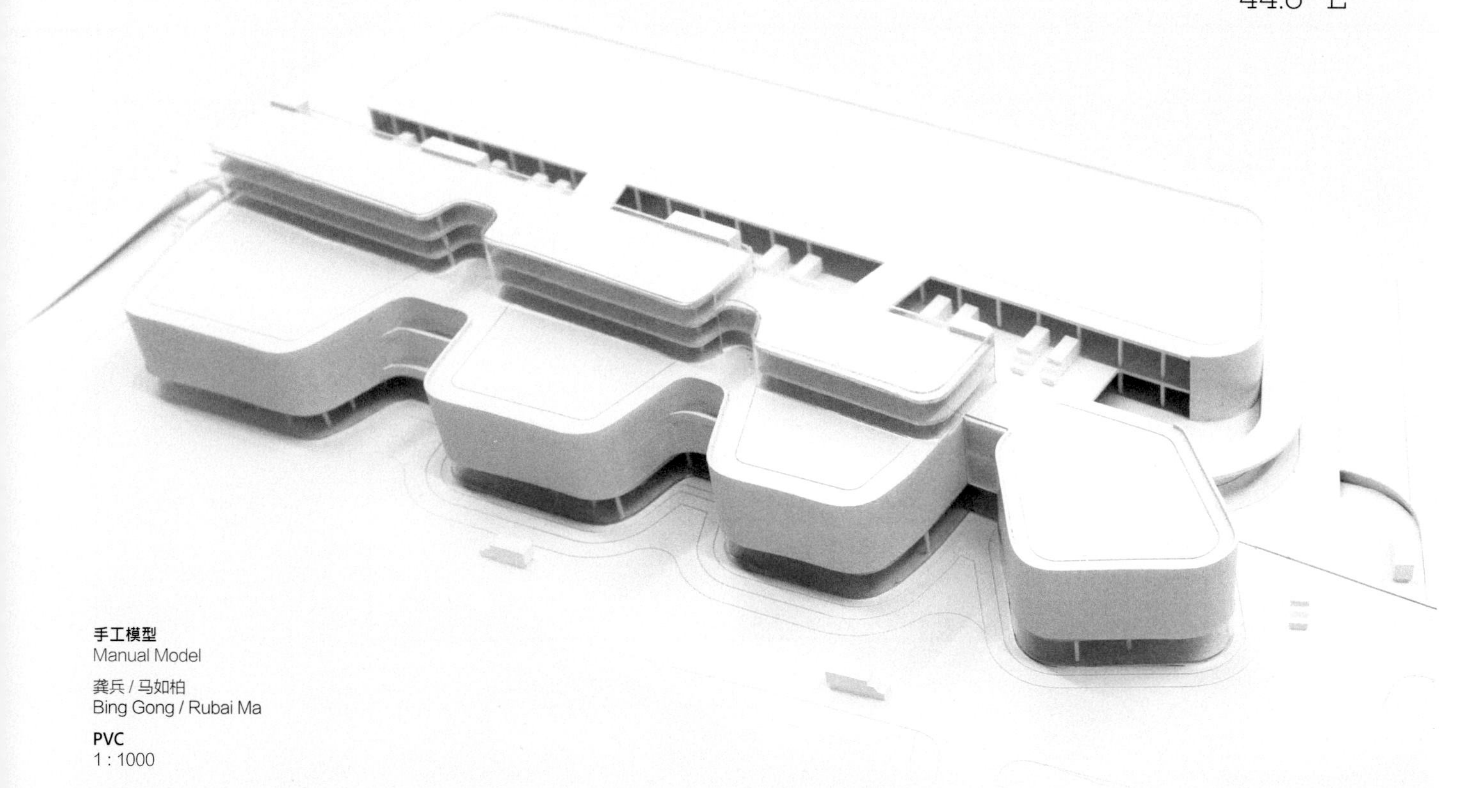

手工模型
Manual Model

龚兵 / 马如柏
Bing Gong / Rubai Ma

PVC
1 : 1000

上海西虹桥、冷链物流中心、辐射长三角
区域配送中心、光明食品集团旗下、物流产业的综合园区
集食品安全及智能物流为一体的全产业冷链综合物流园

Shanghai West Hongqiao, cold chain logistics center, radiant Yangtze River Delta Regional distribution center, Guangming Food Group, logistics industry comprehensive park 、Whole industry cold chain integrated logistics park integrated with food safety and intelligent logistics

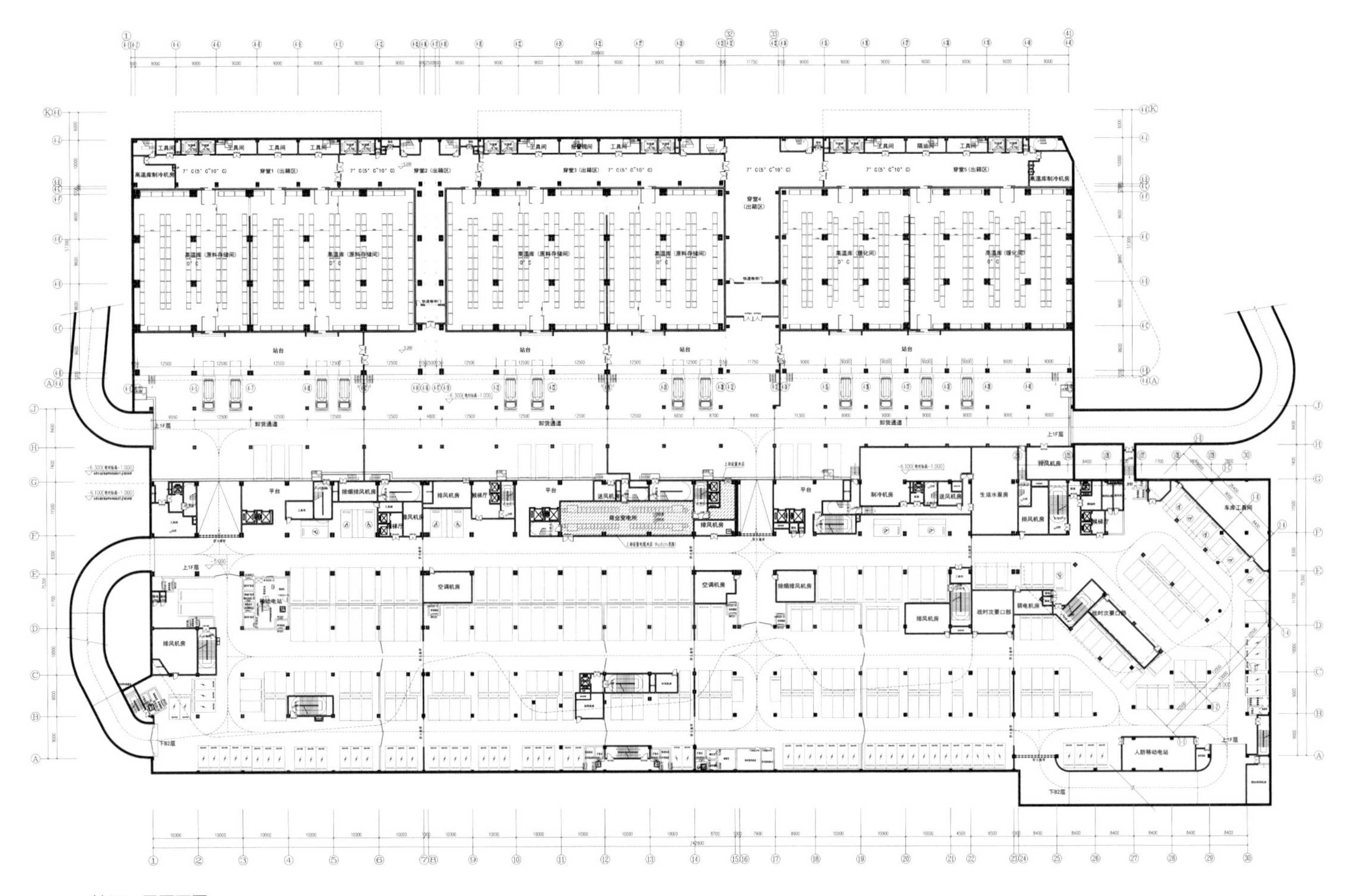

地下一层平面图
Basement Floor Plan

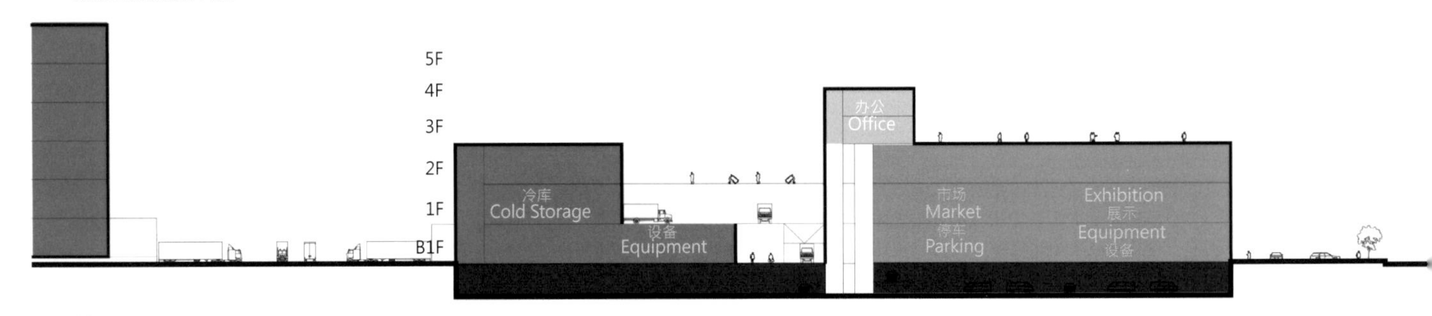

剖面图
Sectional View

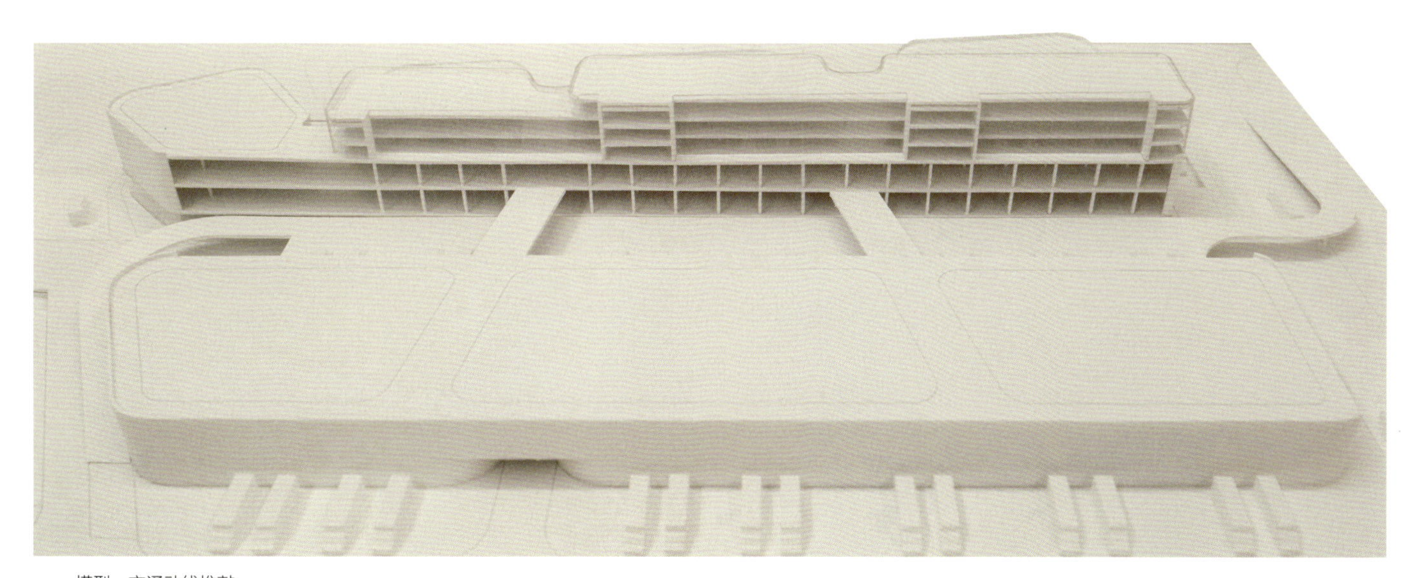

模型：交通动线推敲
Model:Traffic Lines Scrutiny

紧跟物流革新 · 创造新型规划模式

结合策划建议，参考从日本筑地市场到丰州市场的历史沿革，我们看到了现代物流模式的革新与高效特质，并感受到国内现存 B 端市场建筑的不足。经多方案比较探讨，确立了最终规划模式——将冷库用房置于北侧，与一期形成完整的物流园区，便于统一管理和流线组织。将市场置于南侧，背侧与物流用房相连，正面临近城市道路，分为四大模块一字阵列，灵活互换，并延伸最大商业价值面。办公嫁植在市场上，享受 20 000 ㎡的屋顶花园。

在交通流线上，考虑到集卡、小型卡车（2 吨容量以下）和客人小型车辆的三种车流的高效周转，以相对独立的分区和流线设计。园区主要车流沿基地外环设置，单入单出提升效率。办公及参观的人车出入口均设于南侧凤星路，与物流出口保持较远距离，避免流线干扰。沿凤星路铺设的广场延伸至市场底层灰空间，以绿化点缀市场门口，改善传统较为混乱的市场形象。

在与相关专业各部门配合中，我们从实际出发，多维度论证各种大胆的想法，同时将体验型商业设计理念整合入物流园设计中，使之成为新型综合园区的品质标杆。

Keeping Up with Logistics Innovation and Creating a New Planning Model

Combined with the proposal of planning, referring to the historical evolution from Tsukiji Market in Japan to Fengzhou Market, we have seen the innovation and high-efficiency characteristics of the modern model of logistics and felt deficiencies in the existing domestic construction of B-side market. After a comparative study of the plan, the final model of planning was established - putting the room of cold storage on the north, forming a complete logistics park with the first phase to facilitate unified management and organization of streamline. The market is located on the south. The north is connected with the house of logistics. The front is close to the city road. It is divided into four arrays of one word that are flexible exchange, extending the maximum commercial value. Value of Offices enjoy 20,000 square meters of roof garden in the market.

In the streamline of traffic, considering efficient turnaround of the three kinds of traffic flow of trucks, small trucks (The capacity is less than 2 tons) and small vehicles of guest, it is designed as relatively independent partitions and streamlines. The main traffic flow in the park is set along outer ring of the base station. Single entrance and exit improve efficiency. The office entrances and exits of office, visited people and vehicles are located on the south side of Fengxing Road that are kept at a relatively long distance from the outlets of logistics to avoid interference with streamlines. The square paved along Fengxing Road extends to the grey space of ground floor of the market, greening the entrance and improving traditionally confusing image.

Cooperating with various departments of related professions, we demonstrate various bold ideas in multiple dimensions from reality. Meanwhile, we integrate the idea of experience-based commercial design into the design of logistics park, making it a benchmark for quality of the new and comprehensive park.

新型物流形态：专业型冷链园区
New Logistics Form: Professional Cold Chain Park

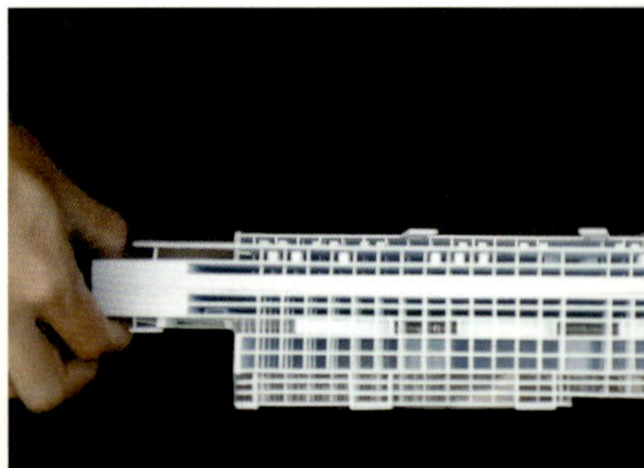

垂直叠加设计与体量推敲
Vertical Superposition of Design and Mass Scruti

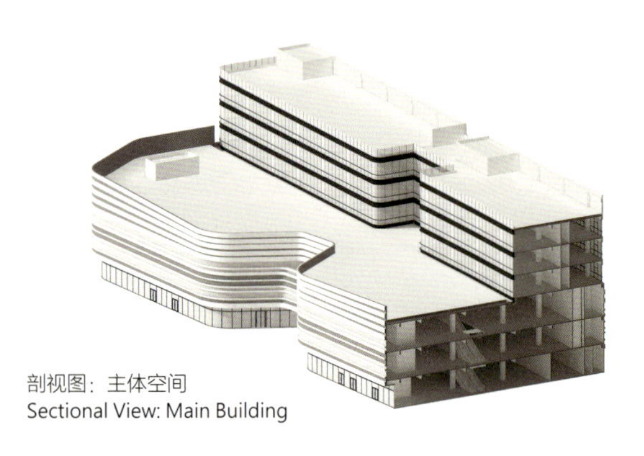

剖视图：主体空间
Sectional View: Main Building

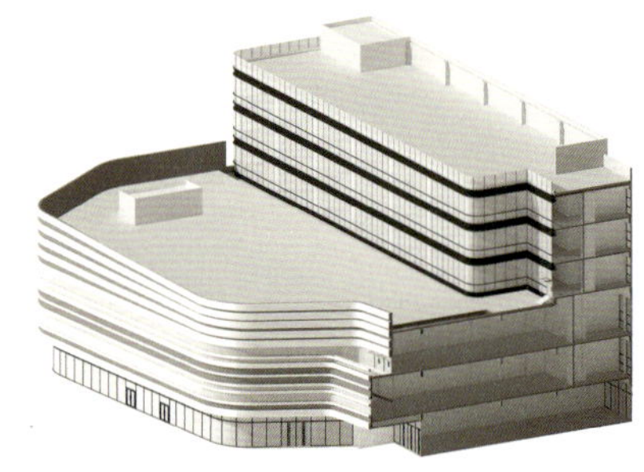

剖视图：中庭空间
Sectional View: Atrium

模块分区设计
Modules Partition Design

10.5m 跨度空间推敲
10.5m Span Space Scrutiny

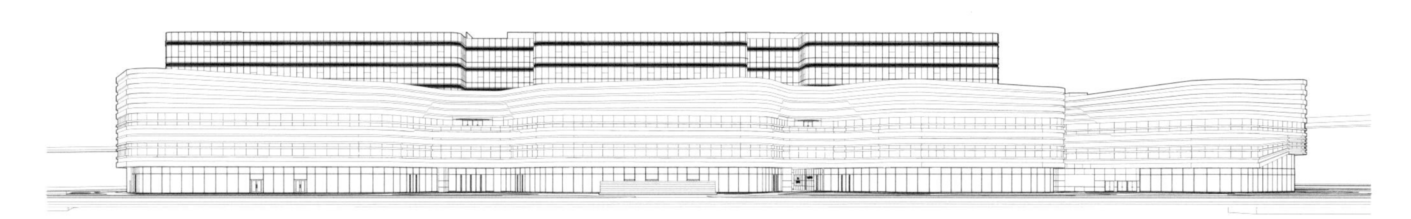

沿凤星路立面
Facade along Fengxing Road

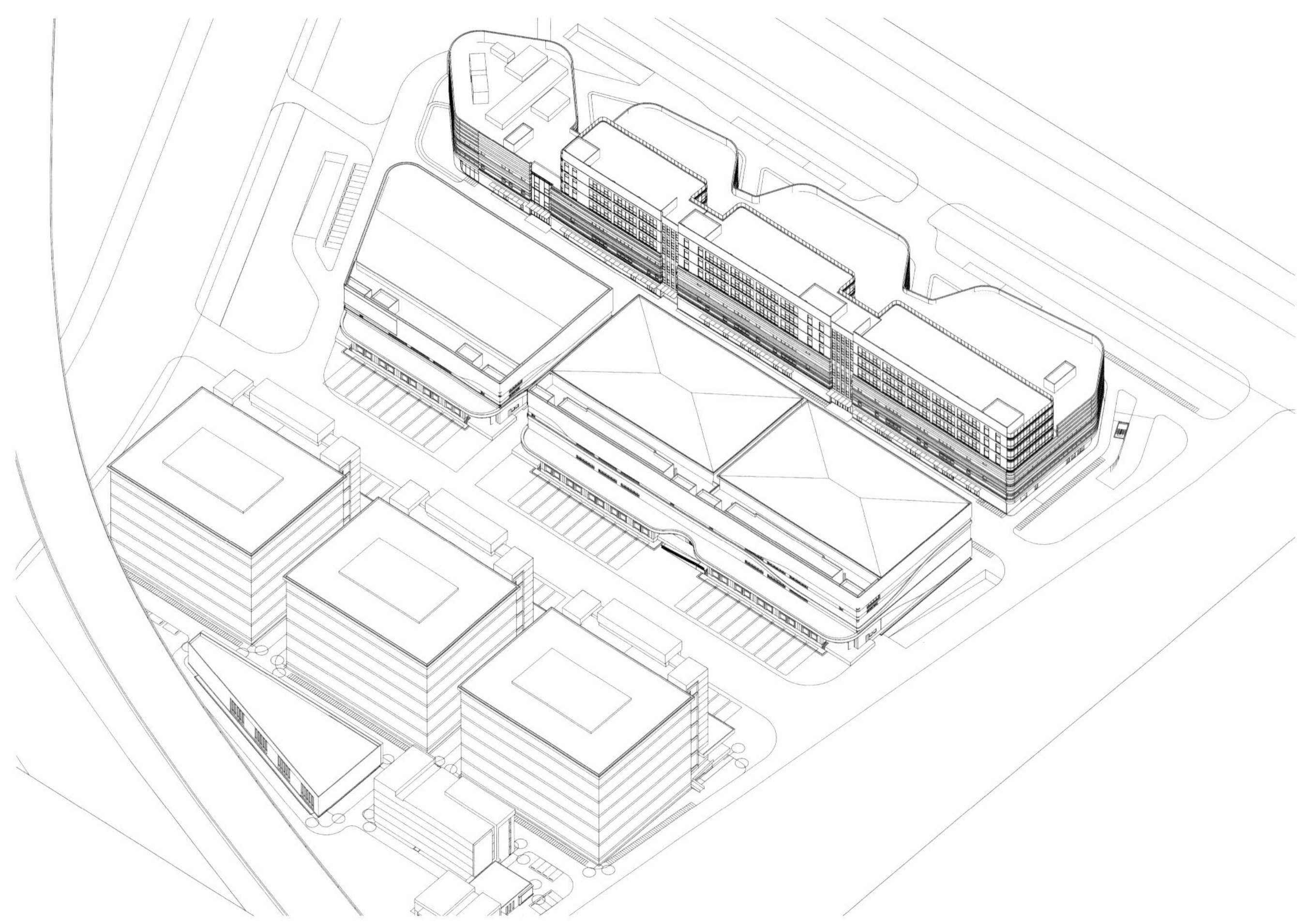

设计关注点：高效流线 > 形态 > 布局
Design Focus: Efficient Lines > Form > Layout

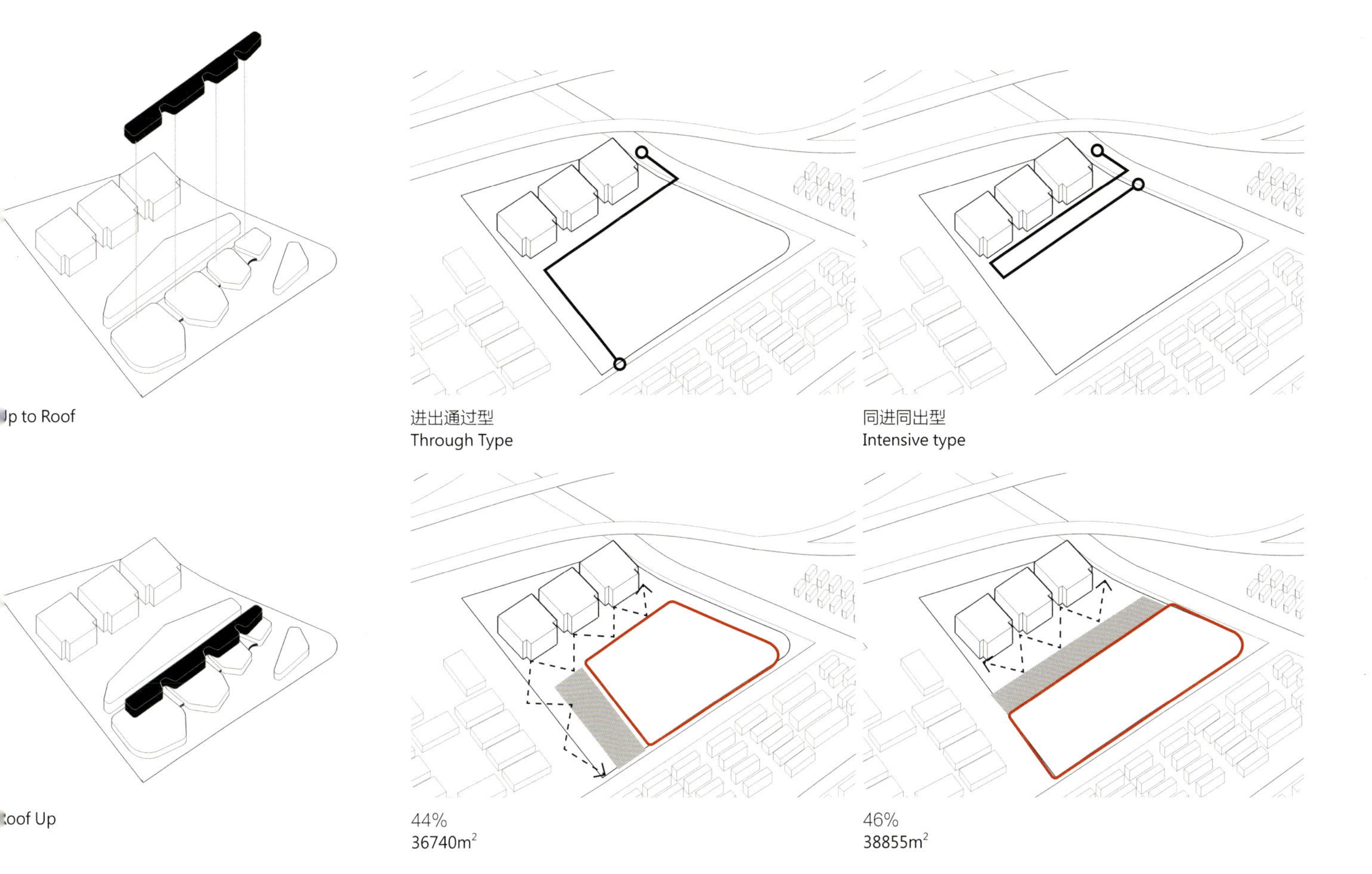

p to Roof

进出通过型
Through Type

同进同出型
Intensive type

oof Up

44%
36740m^2

46%
38855m^2

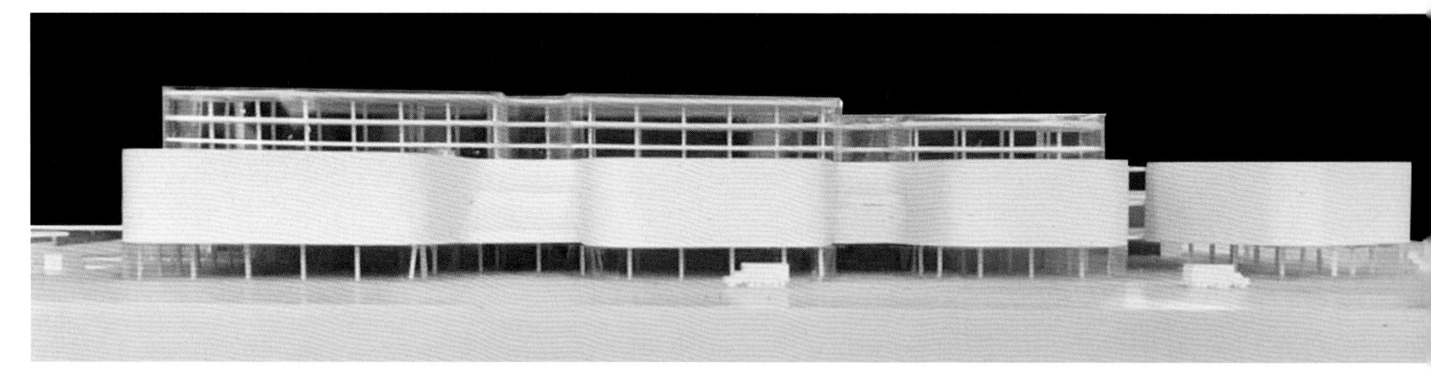

正面：会展区域
Front: Exhibition Area

背面：仓储区域
Back: Storage Area

航空视角，合成鸟瞰
Aviation Perspective, Synthetic Aerial

适应性设计 / 上海南翔小绵羊总部产业园

Adaptive Design / Industrial Park of Nanxiang Small Sheep Headquarter in Shanghai

上海

占地面积 ：13 019 平方米（20 亩）

建筑面积 ：58 864 平方米

容 积 率 ：4.52

设计时间 ：2017

“房地产开发的核心基础资源就是土地。”

改造设计如何突破新旧的界限？

该项目原先为小绵羊的总部厂区，基于政策原因[1]，业主希望将其改造为现代产业办公园区。在上位规划中，项目距离 11 号线陈翔路地铁站 2km，位于大型居住片区与银南翔商务区及拓展区[2]交汇处，邻近上海印象城[3]区域位置良好。无论是业主还是政府，盘活这块土地的价值，都是最关键的着眼点。所以自设计初始，我们从园区运营视野及周边类似案例研究入手，设定了融合高端办公、品牌展示发布、休闲商业、会议、人才公寓的主要业态及配比，强化开放性、生态性的多元复合。基于绿地大场创新产业中心项目经验，我们将产品线式的设计法应用在本次实践中。

注：

1 基于土地高效利用原则，南翔政府对产业用地每亩年税收有新要求。

2 拓展区定位以文化产业为特色、公共服务为基础的区域公共活动中心及综合功能片区。

3 上海印象城为超大型商业综合体。总建筑面积达 32 万平方米，扣除 3 千余车位的巨型停车场及仓库等辅助面积后，B2 至 6 楼的商业楼层面积仍高达 19 万平方米。初步规划 6 楼至负 2 楼分别规划为闲适、娱乐、聚会、家庭、时尚、品质、市集、生活八大主题。其中，零售时尚占比最高，约 32%，餐饮场景占约 25%，生活社交占比约 18%，娱乐文化占比约 9%，家庭亲子占比约 9%，健康休闲占约 7%。

Shanghai

Floor Space : 13,019 m^2

Gross Floor Area : 58,864 m^2

Plot Ratio : 4.52

The Time of Design : 2017

"The Core Basic Resource for Real Estate Development Is Land."

How can the retrofit design break through the old and new boundaries?

The project was originally a factory site for small sheep. Based on policy reasons, the owner hopes to transform it into a modern industrial office park. In the above plan, the project is 2km away from Line 11 Chenxiang Road subway station, and is located at the intersection of large residential area and Yinnanxiang Business District and Development Zone 2. The area near Shanghai Impression City 3 is in good condition. Whether it is the owners or the government, the value of revitalizing this land is the most critical focus. Therefore, starting from the initial stage of the design, we started with a vision of the park's operations and similar case studies in the surrounding areas, and set the main formats and ratios for integrating high-end office, brand display and release, leisure commerce, conferences, and talented apartments, and strengthen the diversity of openness and ecology. complex. Based on the experience of Greenfield Dachang Innovation Industry Center project, we apply the product-line design method in this practice.

Note:

1 Based on the principle of efficient land use, the Nanxiang government has new requirements for the annual tax revenue per year for industrial land.

2 Development Zones A regional public activity center and a comprehensive functional area, which are based on cultural industries and public services.

3 Shanghai Impression City is a super commercial complex. With a total construction area of 320,000 square feet, the commercial floor area of the B2 to 6th floor is still as high as 190,000 square meters after deducting the auxiliary parking area of 3,000-odd car parking lots and warehouses. The preliminary planning of the 6th floor to the 2nd floor is planned as the theme of leisure, entertainment, party, family, fashion, quality, bazaar, and life. Among them, retail fashion accounted for the highest proportion, about 32%, catering scenes accounted for about 25%, social networking accounted for about 18%, entertainment culture accounted for about 9%, family parent-child accounted for about 9%, and health and leisure accounted for about 7%.

产品升级土地新价值

Product Upgrading New Value of Land

31°18′
8.2″ N
121°17′
20.4″ E

南翔、城市近郊、传统企业转型升级
化零为整最大化激活土地价值
产业升级的浪潮下探索园区改造新趋势

Nanxiang, Urban Suburbs, Traditional Enterprise Transformation and Upgrade、Zero to maximize the activation of land value
Under the wave of industrial upgrading, explore the new trend of the transformation of the park

小绵羊
SHEEP

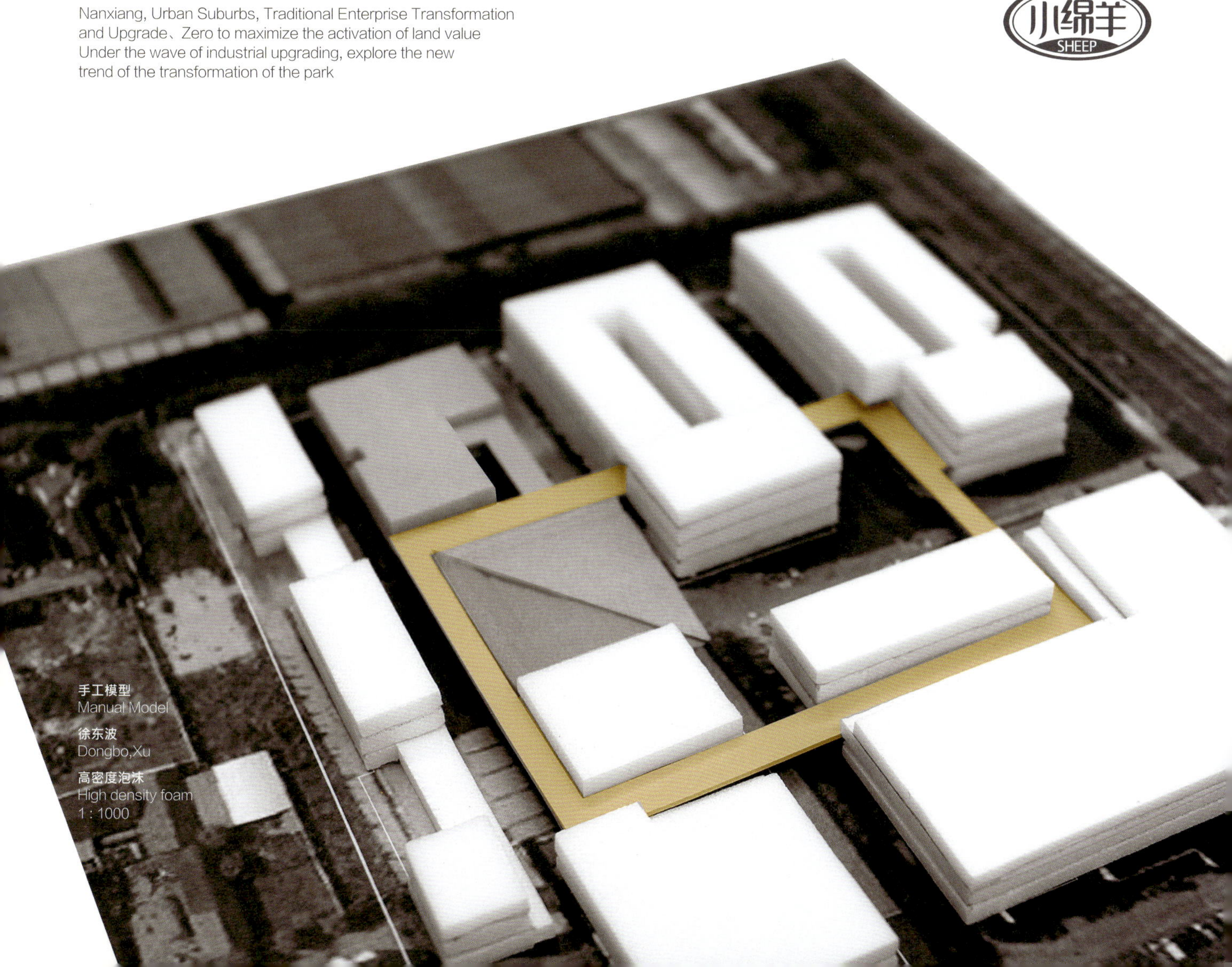

手工模型
Manual Model
徐东波
Dongbo,Xu
高密度泡沫
High density foam
1：1000

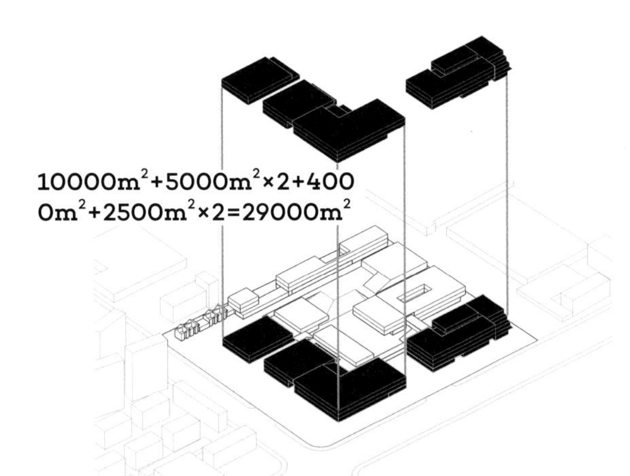

企业总部

企业总部对应街角良好形象昭示性

大中型产品对应园区入口，均拥有良好的展示视野

Correspondence of a good image at the corporate headquarters
Large and medium-sized products correspond to the entrance of the park, all have a good view of the exhibition

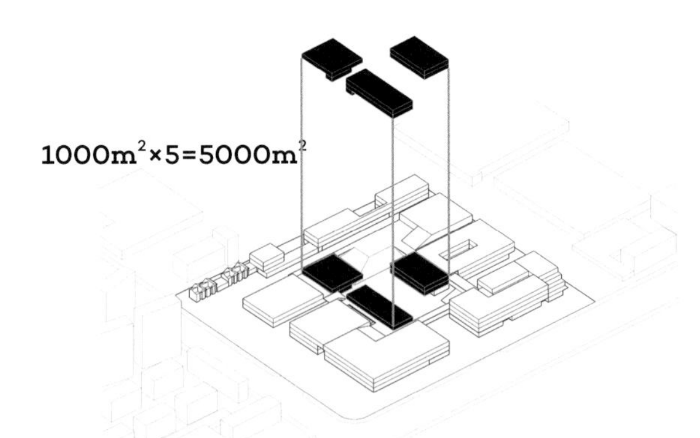

中小型规模企业

从最大体量的四处建筑中切出

1 000 平米产品（中小型规模企业，相对易租）置于园区的核心景观区域

Cut out from the most general mass of buildings 1000 square meters of products (small and medium-sized enterprises, relatively easy to rent) in the park's core landscape area

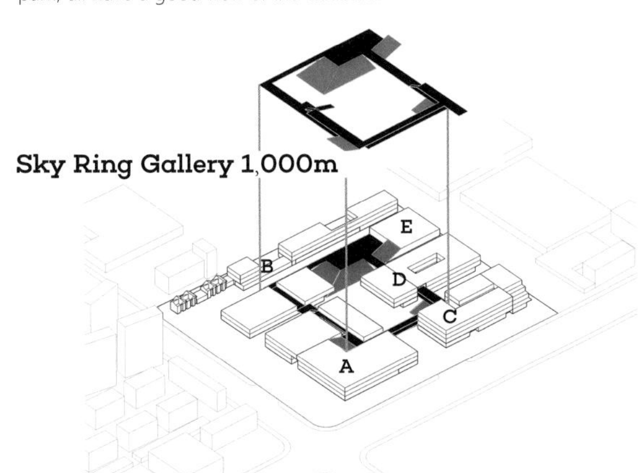

天空云廊

1 000 米二层 “天空云廊” 系统贯穿园区

风雨环廊，人车分流

1,000-meter-second-storey "sky Cloud corridor" system runs through the park
Wind and rain ring corridor, man car shunt

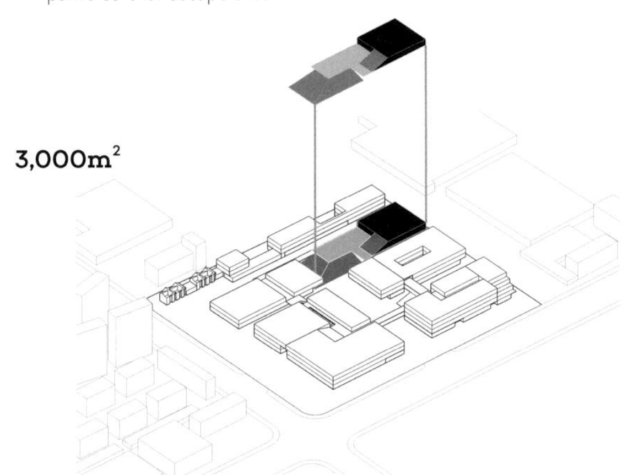

示范交流展览

公益性项目

大型发布展示厅，多功能会议厅

Commonweal Project
Large release showroom, multifunctional Conference room

设计思考

入口小、纵深大的条件下如何激活端部功能?
高密度条件下如何创造丰富趣味的内部空间?
改造项目中“有所不为”设计法的独特意义?

Design Thinking

How to activate the end function under the condition of small inlet and large depth?
How to create an interesting interior space under high density conditions?
What is the unique meaning of the "doing nothing" design method in the transformation project?

315m REACH

Building Density

52.6%

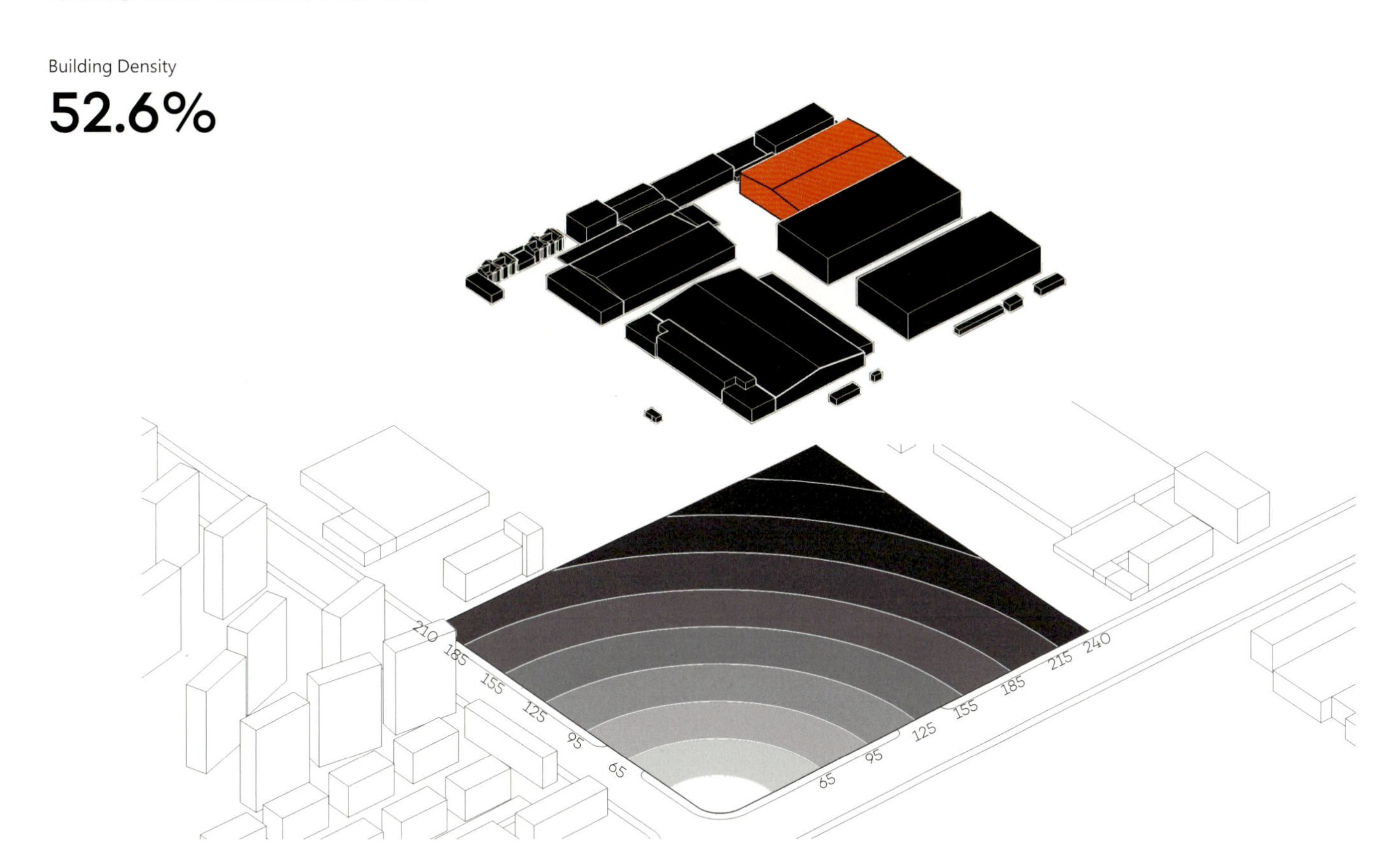

1 中心广场及入口视角
Central Plaza View

3 改造前照片 - 家纺工厂
The Photos before Transformation

2 工作模型
Working Model

如何将规划与产品并行园区运营策略？

规划上，结合周边现状布置不同的产品，引入相应企业，再利用空中环道联系整个园区。按客观昭示性特征，设计将大中型企业置于沿街，创意独栋置于花园中央，小型办公集群于滨河的一个楼里并紧邻公共配套资源。在园区的中心，不仅有大尺度广场作为公共展场和城市客厅，围绕下沉广场还有精致的咖啡外摆和生态绿坡花园。

随着产业升级城市更新的深入，原有的城镇格局与建筑形态需要调整，然而这并不是在一张空白的草图纸上随意涂抹的空想，这点在改造类型的产业园设计中尤为重要。工程复杂度、工期控制、报批流转、招商周期，都会体现在成本控制的压力中，所以改造力度的选择在设计策略中尤为重要。在该项目中，我们最大程度地保留了结构形式，在满足规范要求的情况下，实行改装修流程的更新设计，并配以空中连廊、下沉广场，塑造符合当代气质的产业园。

目前项目正在进一步策划论证中。

How to make the plan and product operate the strategy of park simultaneously?

We arrange different products based on the current status of surrounding areas, introducing relevant companies, using air loop to link the entire park. Based on objective and reprehensible characteristics, large and medium-sized enterprises are placed along the street. Creative single buildings are in the center of the garden. Small office clusters that are close to public supporting resources are located in a building in Binhe River. There are large-scale plazas as public exhibition halls and urban living rooms, and a sublime plaza with exquisite coffee outcrops and ecological gardens of green slope.

With deepening of renewal of industrial upgrading city, original urban structure and forms of architecture need adjusting. However, it is not an arbitrary dream on a blank sketch paper, which is particularly important in the design of industrial parks of the type of renovation. Complexity of engineering, control of duration, approval for circulation and period of attracting investment are reflected in pressure of control of cost. So the choice of strength of transformation is particularly important in strategies. We kept the form of structure to the utmost extent. In the case of meeting the requirements of specification, we implemented the redesign of refitting process with air corridors and sunken squares, creating an industrial park that meets contemporary temperament.

The project is further planed and demonstrated at present.

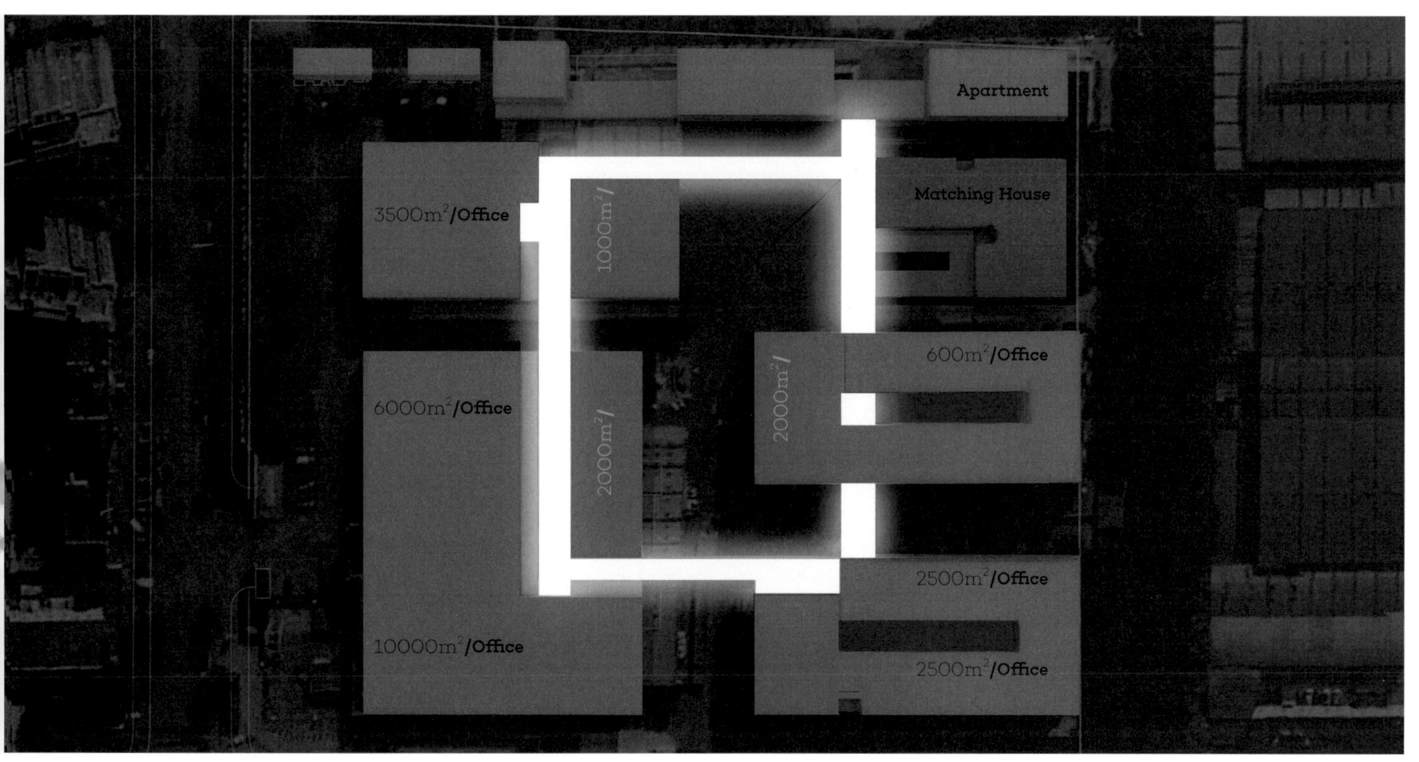

增强联系的广廊与产品分布：第五立面在空港区的昭示作用

Distribution of Corridor and Product of Enhanced Relationship: The Role of the Fifth Facade in the Area of Airport

通过二层链接保持底层地面全开放

Keep the Floor of Bottom Open by Link of the Second Floor

Roof Garden

Roof Parking

600m^{2}Office

Headquarter

600m^{2}Office

Headquarter

600m^{2}Office

Exhibition

Commerce

外科手术式改造 / 上海蓝天数字游戏产业园
Surgical Transformation / Blue Sky Digital Game Industry Park in Shanghai

上海

占地面积　：45 873 平方米

建筑面积　：58 000 平方米

容 积 率　：1.2

设计时间　：2017

地块位于上海嘉定区东南片区，浏翔公路以东，蕴北路以南，占地约 68.8 亩。这是一栋特别的单体建筑改造：单栋四层建筑总量 5.8 万平米，底层 1.8 万平米，如此体量的建筑在上海城区颇为罕见。

该项目前身为上海鸿宝物流信息交易中心，之前有诸如影响周边交通、业态杂乱非法、人流集聚度高且流动频繁、安全隐患高的问题。随着南翔经济的快速发展，该地块迎来政府发展整治该区域的契机，按计划转型开发为游戏创意产业园，由南翔蓝天经济城进行统一规划开发。

项目计划结合周边资源，整体建设打造南翔数字娱乐产业园，成为集游戏、VR、电子竞技等产业为一体的产业链集群，园区将涵盖端游手游研发、VR 技术研发、电竞赛事举办、游戏电竞产品及周边产品展示销售等功能。目前台湾电竞协会、新加坡电竞协会已参与策划，该场馆建成后预计每年将举办逾百次专业电子竞技赛事。

Shanghai

Floor Space　: 45,873 m^2

Gross Floor Area　: 58,000 m^2

Plot Ratio　: 1.2

The Time of Design　: 2017

The plot is located in the southeastern district of Jiading District in Shanghai, east of Liuxiang highway, south of Yunbei Road, covering an area of approximately 68.8 acres. This is a special transformation of single-building : a total of 58,000 square meters of four-story building of single-story. The bottom is 18,000 square meters. Such a volume of building is very rare in the city proper of Shanghai.

The predecessor of the project is the Hongbao Logistics Information Trading Center of Shanghai. It previously had problems such as influencing the surrounding traffic, illegal business status, high concentration frequent flow of people, and high risks of security. With the rapid development of the economy of Nanxiang, the plot welcomes the opportunity to develop and renovate the region by government. It plans to transform and develop the game as a creative industrial park. Nanxiang Blue Sky Economic City will carry out the unified planning and development.

The project is planed to build the Nanxiang Industry Park of Digital Entertainment integrally in conjunction with the surrounding resources, becoming an industry chain cluster that integrates games, VR, e-sports and etc. The park will cover research and development of mobile games, research and development of VR technology , holding electric competition, exhibition, sale and other functions of gaming e-sports products and peripheral products. At present, the E-sports Association of Taiwan and Singapore have participated in the planning. After completion of the stadium, it is estimated that more than one hundred professional e-sports events will be held each year.

蓝天经济城持续发展

Sustainable Development of Blue Sky Economic City

31°19′
27.6″ N
121°19′
8.7″ E

上海近郊、政府主导、巨大的建筑体量、
高度定制化、数字娱乐产业、新时代的消费娱乐需求
建筑与产业共生

In the suburbs of Shanghai, government-led, huge mass of building, highly customized, digital industry of entertainment, symbiosis of buildings and industries of the demand of consumer and entertainment in the new era

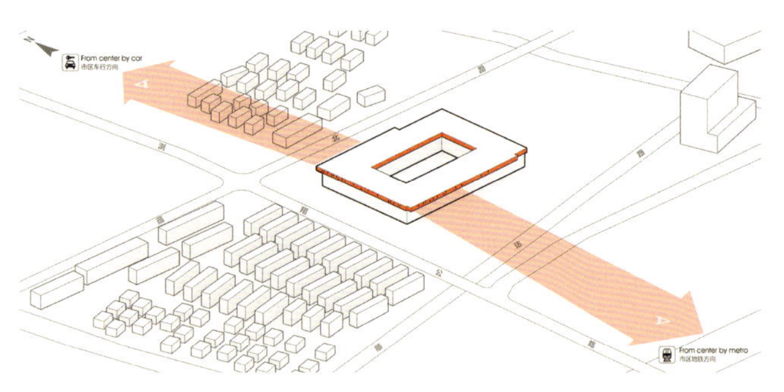

Step1 面向市区来向车流及地铁站方向形成城市界面
Form a Interface to Primary Traffic and Metro Station

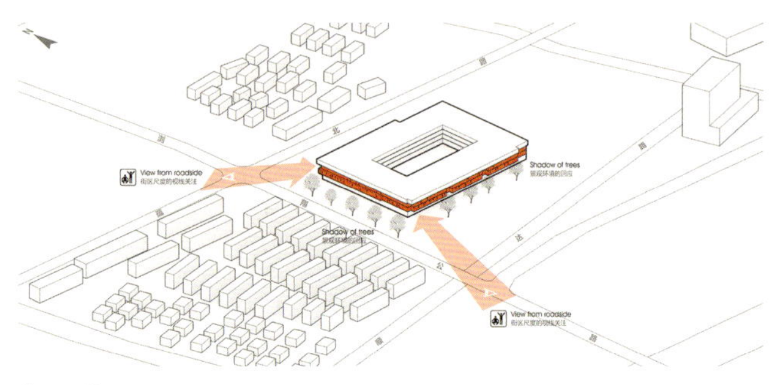

Step2 通过二至三层的折面玻璃形成与环境的对话
Form a Dialogue with the Environment Through Folding Glass from 2F & 3F

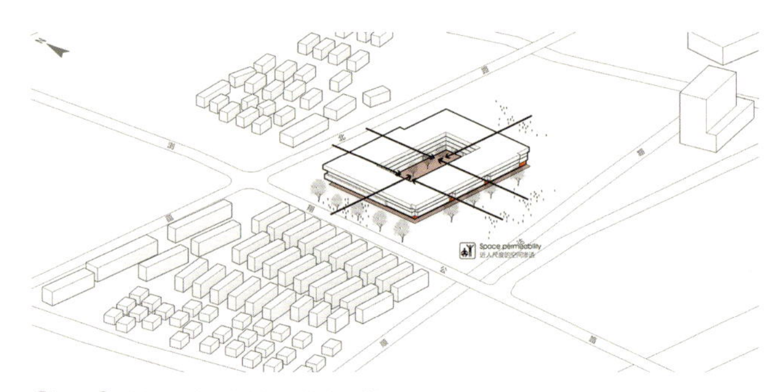

Step3 首层形成回游空间 · 将中庭激活
Activate the Atrium Space through the Grey Space of 1F

建筑外部景观建造条件优越；建筑巨大体量足以创造令人难忘的建筑印象；框架结构形式完好，改造条件优秀；层高具备丰富空间设计可行性。在功能设计上，利用一层局部开放部分回廊，打造首层的洄游空间，让人便利地进入中庭的活动广场，顶层利用局部空间打造客房区域，满足电竞中心综合功能的完整性。

立面设计上采用了菱形玻璃幕墙及通透的幕墙光带，形成了具有地标性的展示建筑，具有科技感的 360° 全景幕墙带，在夜晚可以增强电竞功能及其他数字科技的外延展示性。

The external landscape of the building is in excellent condition;the building is large enough to create an unforgettable impression of architecture. The structure of frame is good and the conditions of renovation are excellent. The floor height has the feasibility of the design of rich space. In terms of design of function, we partially open corridors by a floor to create the first-floor travel space, which allows people to easily enter the event plaza of atrium. The top floor uses local space to create areas of guest room to meet the integrity of the integrated functions of the e-sports center.

The design of facade adopts a curtain wall of diamond glass and a transparent light band of curtain wall, which forms a landmark display building. The 360-degree panoramic curtain wall with a sense of science and technology can enhance the function of eSports and the exhibition of the epitaxy of other digital technologies at night.

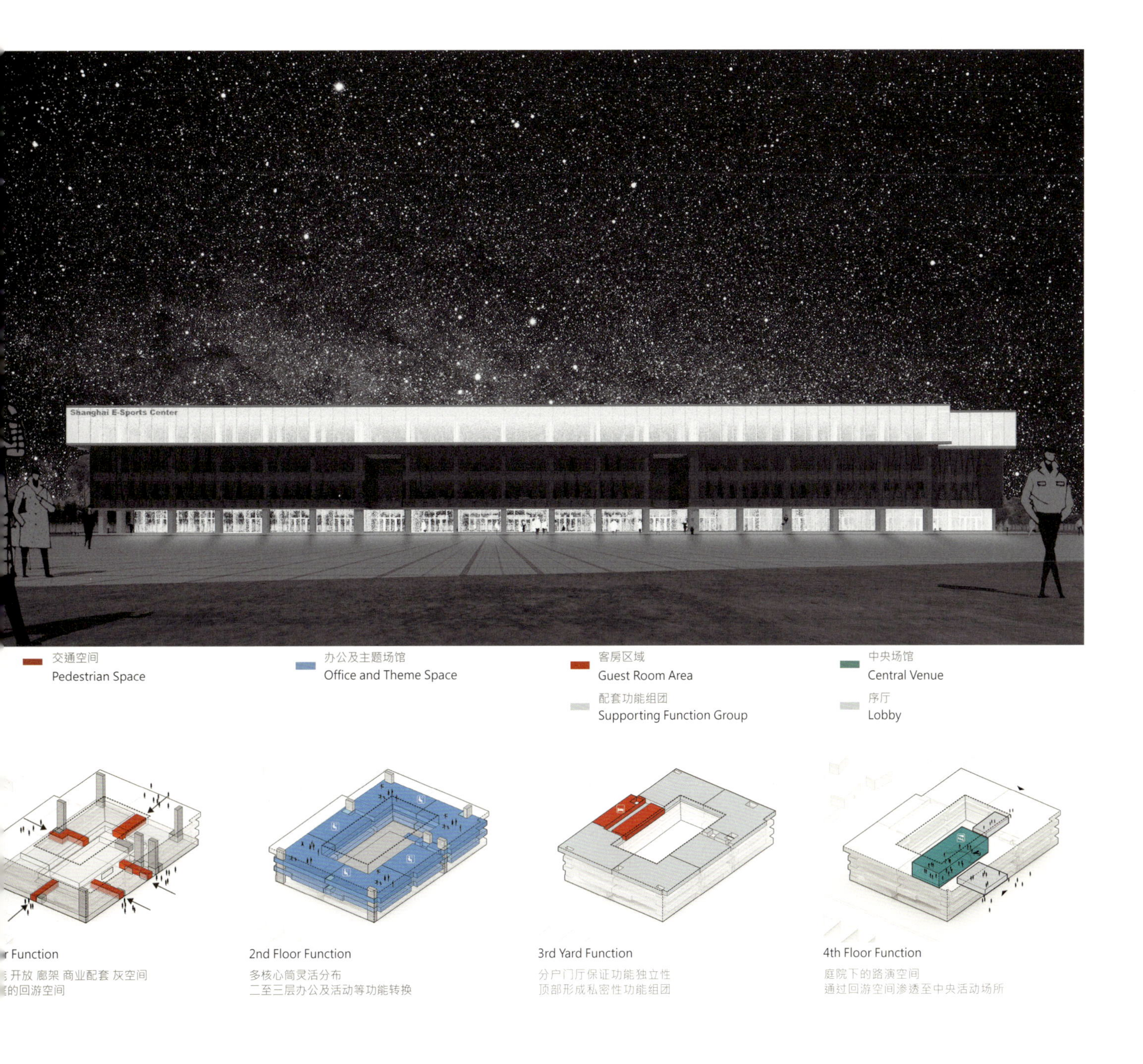

r Function

开放 廊架 商业配套 灰空间
的回游空间

2nd Floor Function

多核心筒灵活分布
二至三层办公及活动等功能转换

3rd Yard Function

分户门厅保证功能独立性
顶部形成私密性功能组团

4th Floor Function

庭院下的路演空间
通过回游空间渗透至中央活动场所

谦虚的设计
Modest Design

第一次工业革命为世界带来巨大财富的同时，机械的模板化生产也与手工反复打磨的精美不断冲突，由此诞生的便是现代设计。从这段起源来看，设计有一项重大的任务，就是平衡生产的效率与审美的需求。

人们需要设计是出于感性的本能——希望有用着顺手看着顺眼的东西，希望过上良好的生活并在其中感受喜悦。但在信息流爆炸的当下，这种欲求却时常被扭曲成铺天盖地的消费心理，设计语言在被消费的用户情景中越来越浮夸，越来越喧哗。欲求的过剩造成设计的轻浮，让人的感性与周围与生活都越来越陌生，而成一试图还原设计的本质，让人更多感受周边城市、建筑与日常进行中的生活。希望通过设计或者再设计来还原这个世界的风貌，在现代设计简单简洁的本质中引起共鸣。

日常原本就很简单。如果每一间房子、每一处场所、每一个物件都如同热情洋溢的演讲，那生活就太喧闹太辛苦了。留给使用者一点宁静的感受空间，让设计以一种谦逊的姿态滋润生活。

3

Galerija
Gallery
ギャラリー

远道 [水流无形 · 随势姿丽]
A long way " The flow of water is invisible, with the posture and beauty "

张佾媛
Yiyuan，Zhang

镜片，麻纸，水墨、碳
Lenses，Hemp paper, Ink, Carbon
原尺寸 Original size 33mm × 33mm
2018

应对变化的节奏 / 南京珠江路创客大街改造

Coping with the Changing Rhythm/ The Renovation in Chuangke Street, Zhujiang Road, Nanjing

南京

占地面积 ： 7 474 平方米（400 亩）

建筑面积 ： 36 120 平方米

容 积 率 ： 4.8

幕墙面积 ： 16 000 平方米

设计时间 ： 2017

Nanjing

Floor Space : 7,474 m^2

Gross Floor Area : 36,120 m^2

Plot Ratio : 4.8

Curtain Wall Area : 16,000 m^2

The Time of Design : 2017

南京重点系统工程

Project of Key System in Nanjing

电子行业曾有句名言“北有中关村，南有珠江路。”

A famous saying in electronics industry is: "There is Zhongguancun in the north and Zhujiang Road in the south."

特色且特殊的项目背景

Characteristic and Special Background of The Project

在南京，几乎没有人不知道珠江路。在国内，这条路也是名声显赫。“北有中关村，南有珠江路”，这是一个阶段国内电子科技行业界的共识。珠江路的电脑价格，可以影响甚至左右整个华东地区的电脑市场。

Almost no one doesn't know Zhujiang Road in Nanjing. In China, it is also famous. The famous saying is a consensus of domestic industry of electron and technology at a stage. The price of computer in Zhujiang Road can affect or even control the entire market of computer in Eastern China.

在互联网时代，一个互联网产品的聚集地，总是充满了熙熙攘攘的“时代气息”。但不可预知的是，电商来了，珠江路原有的价格优势很快被蚕食殆尽，于是，不可避免的萧条来了，转型是必然的选择。

In the age of Internet, a gathering place for Internet products is always full of bustling "epochal atmosphere". But it is unpredictable that the original advantage of price of Zhujiang Road will soon be eroded with e-commerce. So inevitable depression comes. Transformation is an inevitable choice.

未来城，位于珠江路的核心地段。地铁上盖、正对区政府、秦淮河畔、四百米城市主面，优势资源的背景下，近六万平方米的体量对珠江路的发展及转型有着不容忽视的影响力。然而，未来城现状产业杂乱、外观老旧，甚至 1/3 荒废，强烈的资源落差，使政府决定将其列为首轮改造重点。资源越好、越受注目的项目，往往设计难度越高，高关注度下各方条件的谨慎权衡，而改造项目因旧建筑条件的诸多不确定性，设计更需要精力。然而，未来城改造最大的难度，却来自时间。由于被 5 月 20 日的江苏省发展大会选为观摩案例，从项目启动，到样板段施工完成的时间仅 59 天，左手任务右手“军令状”，双手的挑战，来得突然而猛烈。

Future City is located in the heart of Zhujiang Road. Under the background of advantageous resources of the cover of subway, facing the district government, Qinhuai River and the main surface of 400-meter city, the volume of nearly 60,000 square meters has an indispensable influence on the development and transformation of Zhujiang Road. But the current status of Future City is in a mess, old-fashioned, even 1/3-abandoned and a strong gap of resources, which makes the government decide to list it as the first round of reform. The better resources, more attention is paid to projects, more difficulties are in design. The cautious trade-offs between the conditions of parties under high attention and many uncertainties of the project of reforming because of conditions of old construction require design more energy. But the greatest difficulty comes from time. Due to the selection of Development Conference of Jiangsu Province on May 20 as a case of observation, it costed only 59 days from the start of the project to the construction of the model section. The task is in the left hand and "military order" is in the right. The challenge of both hands was sudden and violent.

有限选择下的谦虚设计

Modest Design Under the Limited Options

改造项目需要综合考虑城市环境、文脉传承、改造技术、实施难度等因素。而考虑到工艺、供货、安装、运输等客观时间需求，在材料、构造、手法方面，设计的选择相当有限，如何有策略的做出优秀的方案？

The transformation of projects needs to take into consideration urban environment, cultural heritage, transformation of technologies and difficulties of implementation. Considering requirements of objective time for process, supply, installation and transportation, the choices of design are rather limited in terms of materials, construction and techniques. How can we make good plans strategically?

融于环境的整体打造：地块的最主要环境特征，是沿珠江路的数十棵近七十年法国梧桐，怎样让建筑融入其中是设计的首要考虑点。首开段沿珠江路长约 170 米，设计采用折面玻璃幕墙包裹整个北立面，化解冗长界面单调感的同时，折面可以映衬出更多的法桐倒影与环境肌理，让立面随着天气、四季的变迁改变着样貌，使建筑与城市景观完美地融合在一起。

To integrate into overall environment, the most important environmental characteristics of the plot are dozens of platanus trees along Zhujiang Road for nearly seventy years. The first consideration of design is how to integrate the architecture into it. The first opening section is about 170 meters long along it. The design adopts a folding glass curtain wall to wrap the entire northern facade. While solving the tedious and monotonous feeling of the interface, the folding surface can reflect more inverted image of platanus tress and environment texture, allowing the facade to change appearance with weather and seasons, letting the building integrate into landscape of the city perfectly.

重要节点的适度突出：沿街角面及入口是最主要的形象节点，通过钢结构设计在此位置置入了两个玻璃体块，形成了视线的聚焦点，我们希望以此做出适度的标志性。

The important nodes are moderately prominent. The edged surface and entrance along the street is the most important node of images. Through the design of steel structure, two vitreous blocks are placed at the position, forming the focus of line of sight. We hope to make a modest landmark.

总统府旁、玄武区政府对面、地铁上盖、秦淮河畔、1912 隔壁、数十棵法国梧桐守护 400 米街长、珠江路核心位置。60 日时间，如何留给城市一个崭新的形象？

Next to the Presidential Office, Opposite to the Government of Xuanwu District, a subway station, Qinhuai River, Next to the 1912, Dozens of French Sycamore Guarding the 400-meter street, The Core of Zhujiang Road.
How to leave a new image to the city in 60 days?

距离总统府 120m
120 Metres From the Presidential Palace

玄武区政府门厅视角
Opposite of the Xuanwu District Government

浮桥地铁站正对面
Opposite of the Pontoon Subway Station

32°02′
59.5″ N
118°47′
35.8″ E

Nanjing Presidential Palace

1912

约 1/3 界面视野 占据中心区珠江路总长度 400 米长
满满的法国梧桐遮挡，来自民国时期的“古树”
About 1/3 of the Interface View Occupies 400 Meters of the Total Length of Zhujiang Road in Central Area.
Full of Plane, from the "Ancient Tree" of the Republic of China

即将
创业大街
敬请期待
创业大街
激情永不止步
浮 桥

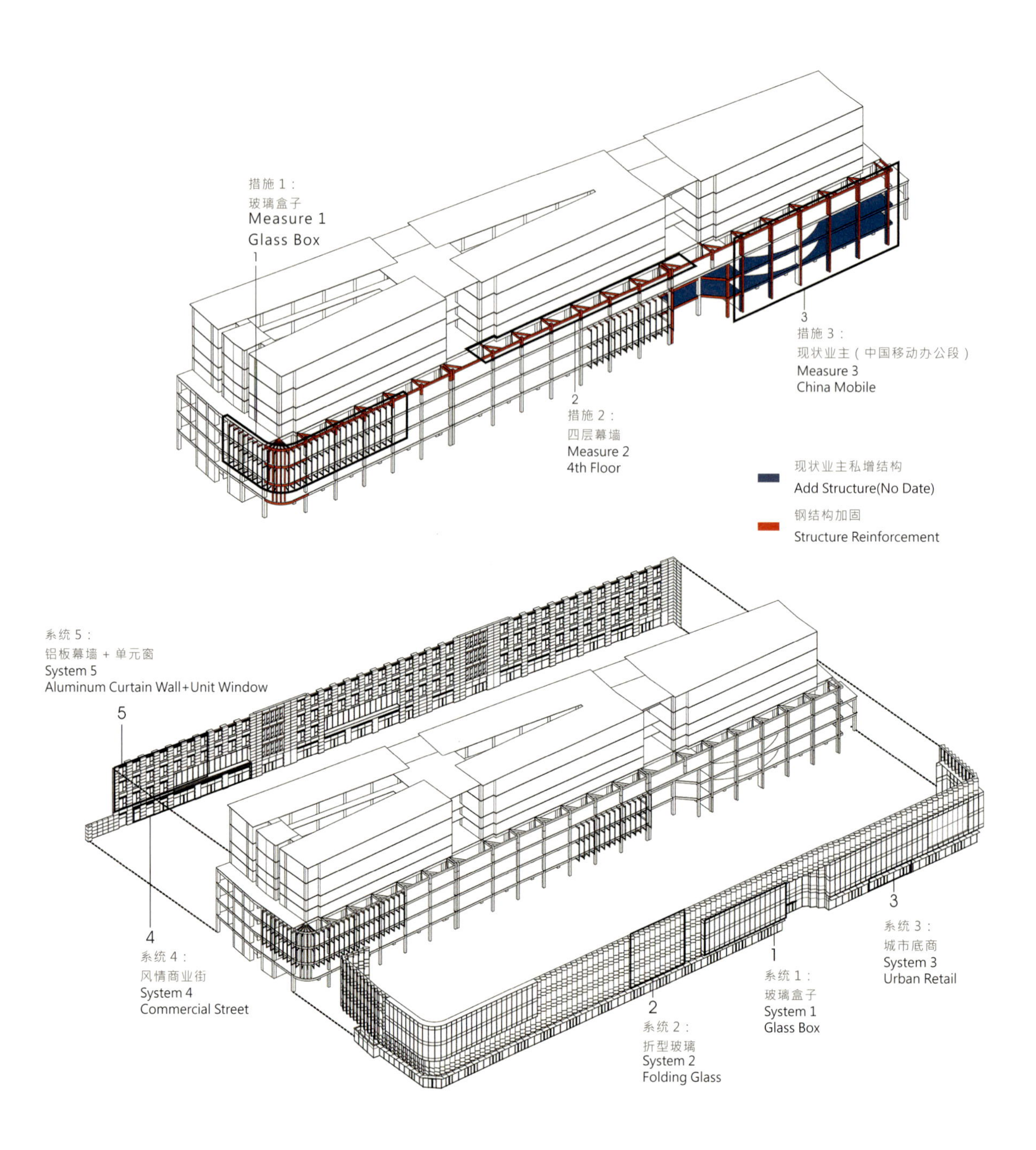

措施 1：
玻璃盒子
Measure 1
Glass Box
1
2
措施 2：
四层幕墙
Measure 2
4th Floor
3
措施 3：
现状业主（中国移动办公段）
Measure 3
China Mobile
现状业主私增结构
Add Structure(No Date)
钢结构加固
Structure Reinforcement
系统 5：
铝板幕墙 + 单元窗
System 5
Aluminum Curtain Wall+Unit Window
5
4
系统 4：
风情商业街
System 4
Commercial Street
2
系统 2：
折型玻璃
System 2
Folding Glass
1
系统 1：
玻璃盒子
System 1
Glass Box
3
系统 3：
城市底商
System 3
Urban Retail

细节的力量

由于项目施工周期与备料极为紧张，立面采用相对简洁的设计方案。而对于这种设计，细节的把控对于品质的体现则显得尤为重要。例如，为了让折面玻璃幕墙既具整体感，又有层次感，方案在折面交接处飞出一段玻璃，挡住后面的竖框，从而有了更生动的材质观感。又如，为了让街角玻璃盒子有更轻盈光滑的表面，我们设计了大尺寸的弧形超白玻璃，从而让表皮材料更平滑的交接。在部分因故无法更新的立面上，设计采用了竖向条纹彩釉玻璃去遮挡破旧的建筑表皮，经过对彩釉图案的细致推敲，让其焕然一新，形成新的形象亮点。诸如此类，虽然改造时常面对材料加工周期长的超时风险，但为了实现最佳的体验感，我们协调各方努力，在核心设计中，还是坚持呈现了原本方案。

如果说大的原则与框架是设计的灵魂与骨骼，细节就是血肉。未来城项目最终呈现出的优秀品质，正因为细节的雕琢打磨。

无法预知的困难

改造项目中存在着原有结构风险，产权归属混乱，原有业主改造抗性等不确定因素，会对设计方案的实施造成不同程度的影响。由于未来城项目极短的设计周期以及改造项目的特殊性，这些因素必然无法全部排查清楚。面对这些无法预知并随着项目推进不断出现的困难，设计需要根据具体情况随时跟进，在多方顾问的配合下快速提出合理的解决方案。

例如，原建筑部分位置存在原有结构不牢，无法承载幕墙重量的问题，于是方案采用了钢结构予以加固，并通过悬挂钢结构吊柱，解决了街角玻璃盒子大尺度悬挑的问题。启动段西侧被中国移动整体租用，不仅心理上改造抗性较强，而且沿街私建的部分在有一定安全风险。经过与结构顾问的现场勘查与技术沟通，出于降低沟通成本与规避可能的结构风险目的，设计决定北立面完全新加钢结构用以承载幕墙系统，与原私搭钢结构系统完全脱开。移动段采用竖向条纹彩釉玻璃遮挡原有的杂乱立面，与东侧的玻璃盒子相互呼应。

The Power of Details

Because the period of project construction and stock preparation are extremely tight, the facade is adopted a relatively simple scheme of design. For this kind of design, the control of details is particularly important for the embodiment of quality. For example, in order to allow the folding glass wall to have a sense of unity and layering, the program flies out a section of glass at the intersection of the folded surfaces to block the rear mullion, thereby giving a more vivid look to the material. For another example, in order to make the glass box of corner have a lighter and smoother surface, we designed a large-sized and curved ultra-clear glass to allow the skin material to transfer smoothly. In some facades that cannot be renewed for some reason, vertical streaked glazed glass was used in the design to mask the worn-out skin of architecture. After detailed scrutiny of the color glaze pattern, it made a new look and formed a new highlight of image. As such, although transformation often faces the long-term risk of the period of material processing, we coordinate efforts of all parties to persist in presenting the original plan in the core design in order to realize the best sense of experience.

If the big principles and framework are the soul and skeleton of design, details are flesh and blood. The excellent quality that the Future City project ultimately presents is precisely the result of the meticulous polishing of details.

Unpredictable Difficulties

There are risks of the inherent structure in the project of renovation. The disorder of property ownership, the original owner's resistance of transformation and other uncertainties will affect the implementation of the design program in varying degrees. Due to the extremely short period of design and the particularity of the project of renovation of Future City, these factors cannot necessarily be completely and clearly investigated. Facing these unforeseen difficulties and the difficulties that have arisen as the project continues to advance, design needs to be followed up at any time according to the specific circumstances. We quickly propose reasonable solutions with the cooperation of consultants in various fields.

For example, in the part of original building, there was a problem that the original structure was not strong enough to carry the weight of the curtain wall. Therefore, the steel structure was used to reinforce it and it was suspended by hanging the steel structure to solve the large-scale overhang of the glass box on the corner. The west side of the start-up section was integrally leased by China Mobile. Not only did the resistance of psychological transformation is strong, but also there are some risks of security in the part of private construction along the streets. After field investigation and technical communication by the consultants of structure, the design decided that the fully-advanced steel structure on the north facade was used to load the system of curtain wall that is completely decoupled from the original system of private steel structure for the purpose of reducing costs of communication and avoiding possible risks of structure. The moving section uses vertical stripes of glazed glass to cover the original messy facade, echoing with the glass box on the east side.

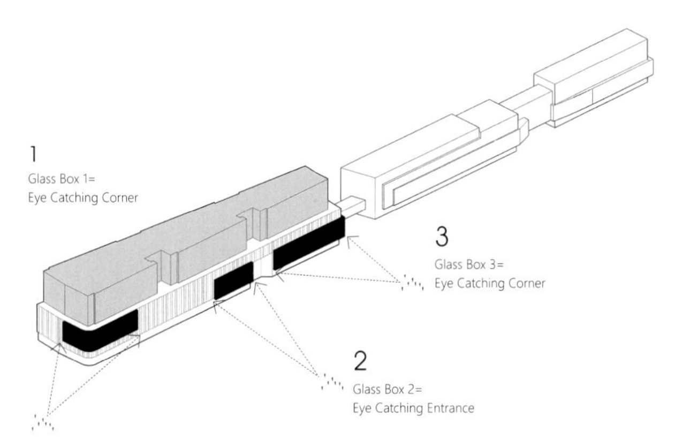

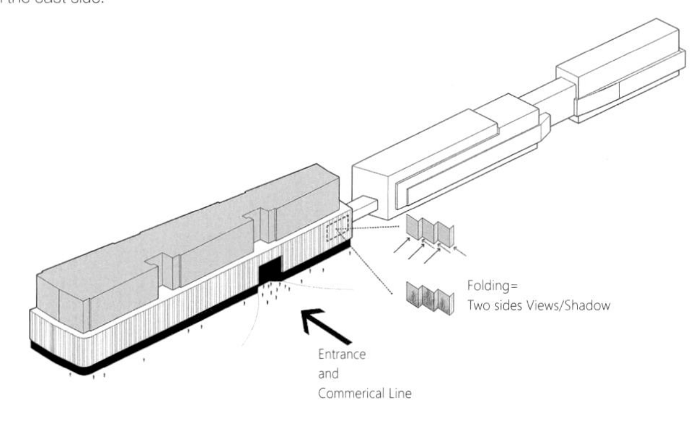

折型玻璃：与树共生，融于城市环境
Folded Glass: Symbiotic with Trees, Integrated into the Urban Environment

与树共生，融于城市环境
Symbiotic with Trees, Integrated into the Urban Environment

公交站台后的整体形象

The Overall Image behind the Bus Station

暴雪后的未来城 B 座照片，摄于 2018 年 1 月 The Photo of B Seat of Future City after Blizzard, Photographed in 2018.1

暴雪后的未来城 B 座照片，摄于 2018 年 1 月 The Photo of B Seat of Future City after Blizzard, Photographed in 2018.1

摄于 2017 年 11 月，秦淮河视角
Photographed in 2017.11, in the Perspective of Qinhuai River

秦淮河视角下的南立面幕墙
South Facade Curtain Wall in the Perspective of the Qinhuai River

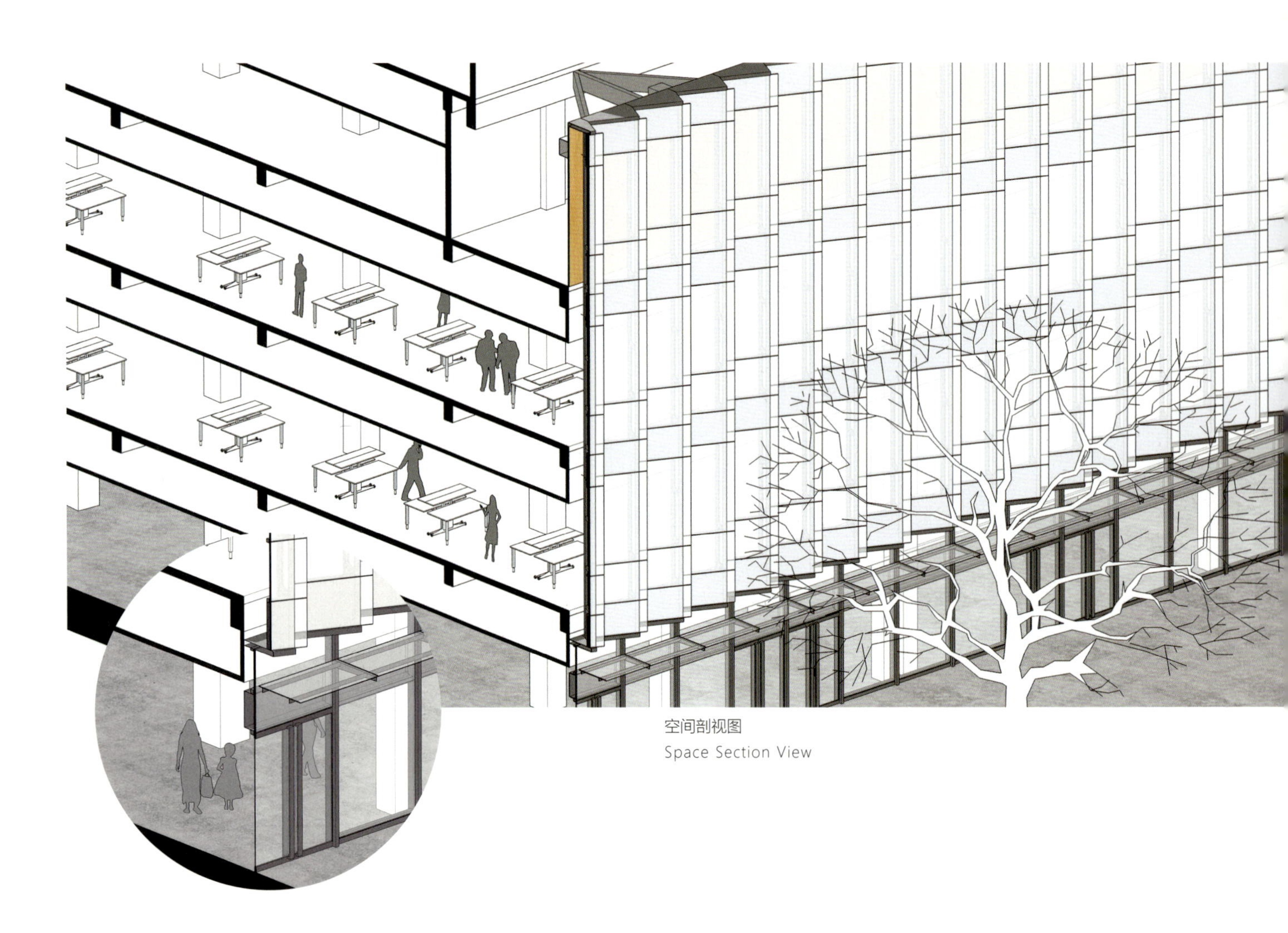

空间剖视图
Space Section View

与树共生

Symbiotic with the Tree

施工进展过程

Construction Progress

珠江路太平桥北，建造过程，摄于 2017 年 11 月

Northern Taiping Bridge in Zhujinag Road, the Process of Construction, Photographed in 2017.11

1. 北立面移动段钢结构加建

Construction of Steel Structure for North Elevation Section

2. 北立面裙房屋顶钢结构加固

Construction of Steel Structure for North Podium Roof

3. 端头钢结构巨构

The End of the Steel Structure Giant

南立面渲染图
The Rendergraph of Southern Facade

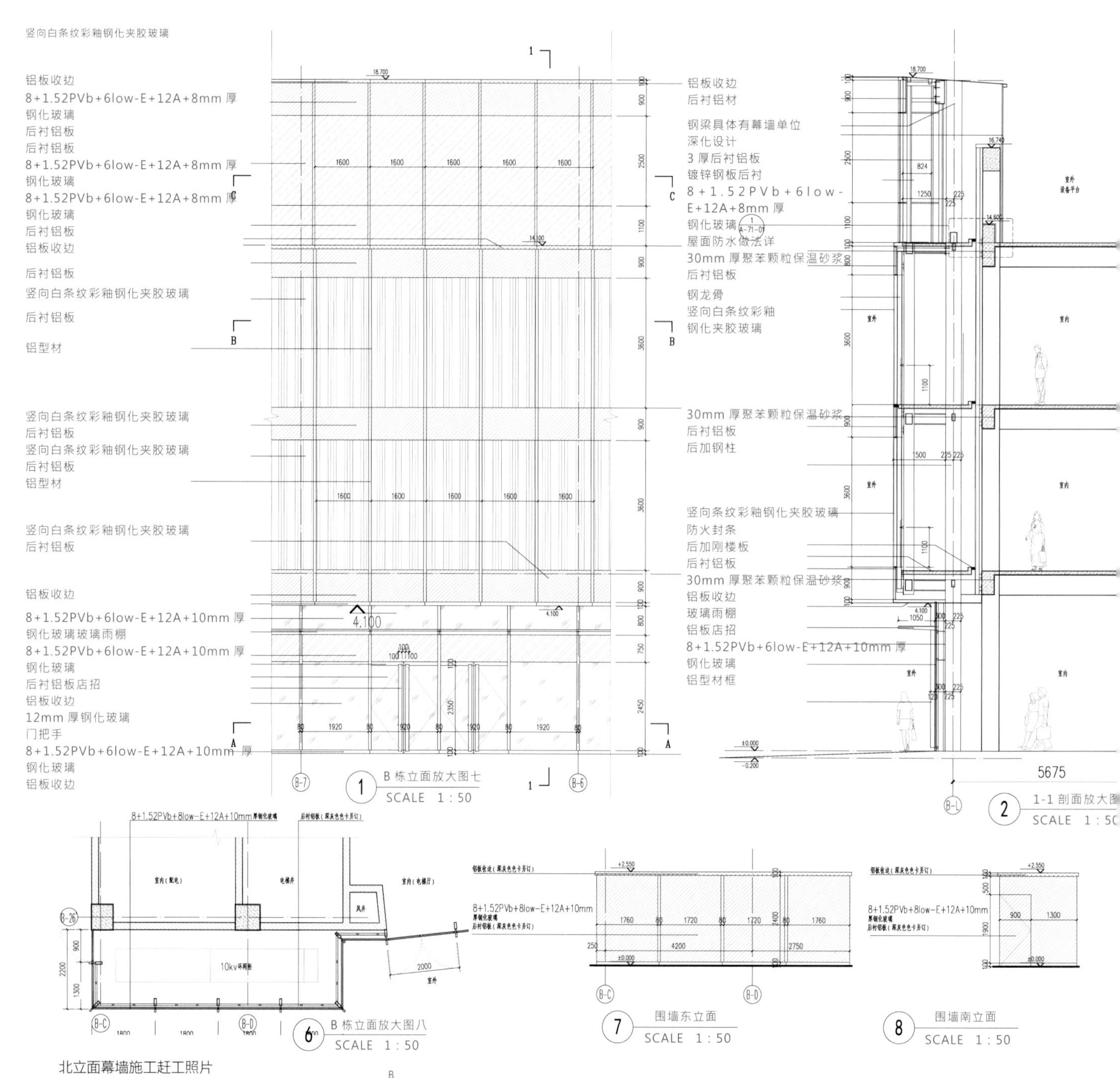

北立面幕墙施工赶工照片

B

The Photo of the Construction and Charrette of the Curtain Wall of Northern Facade

北立面墙身
North Facade Section

南立面墙身
The Body of the Wall of Southern Facade

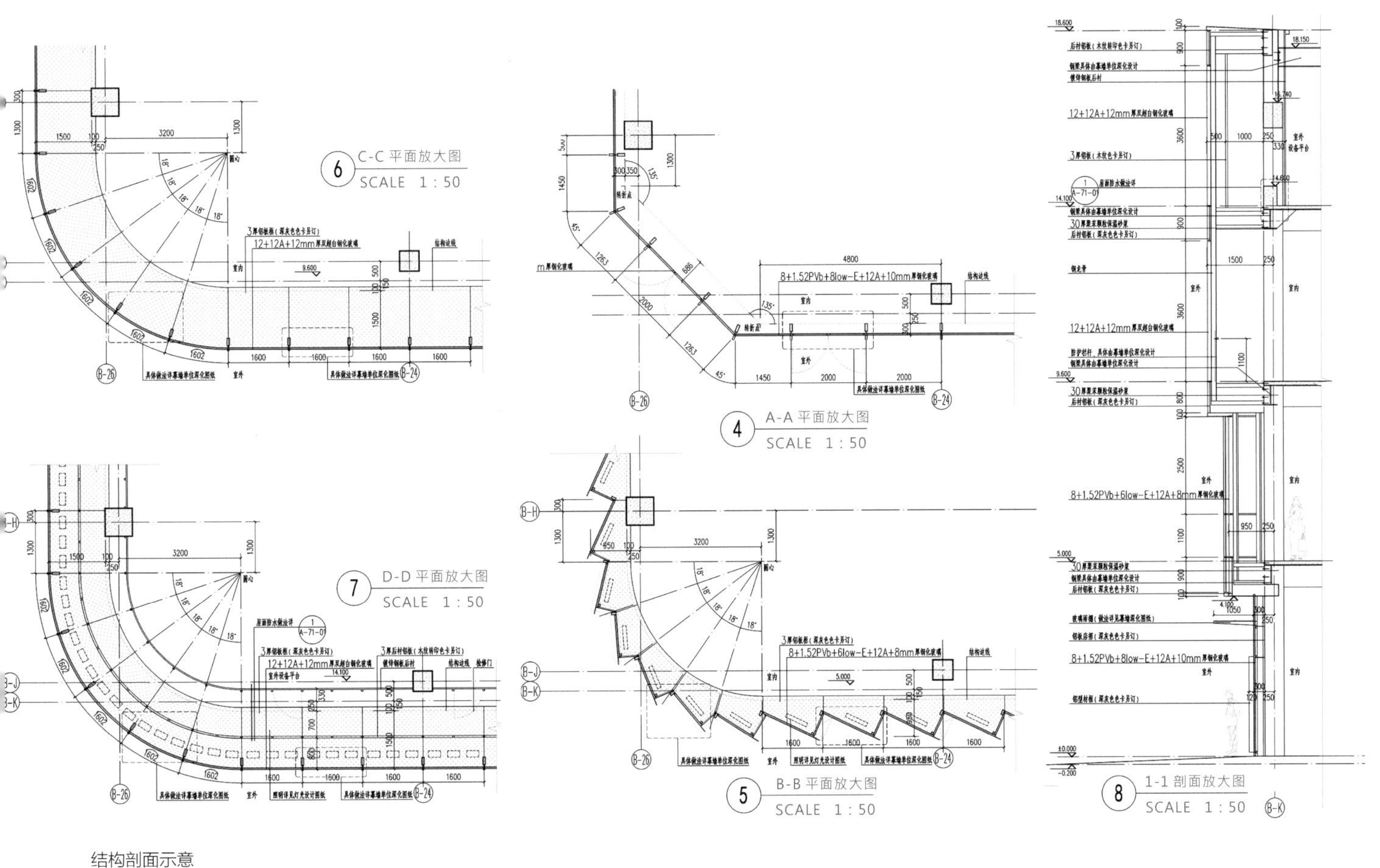

结构剖面示意

Structural Section

体块模型

The Model of the Bulk

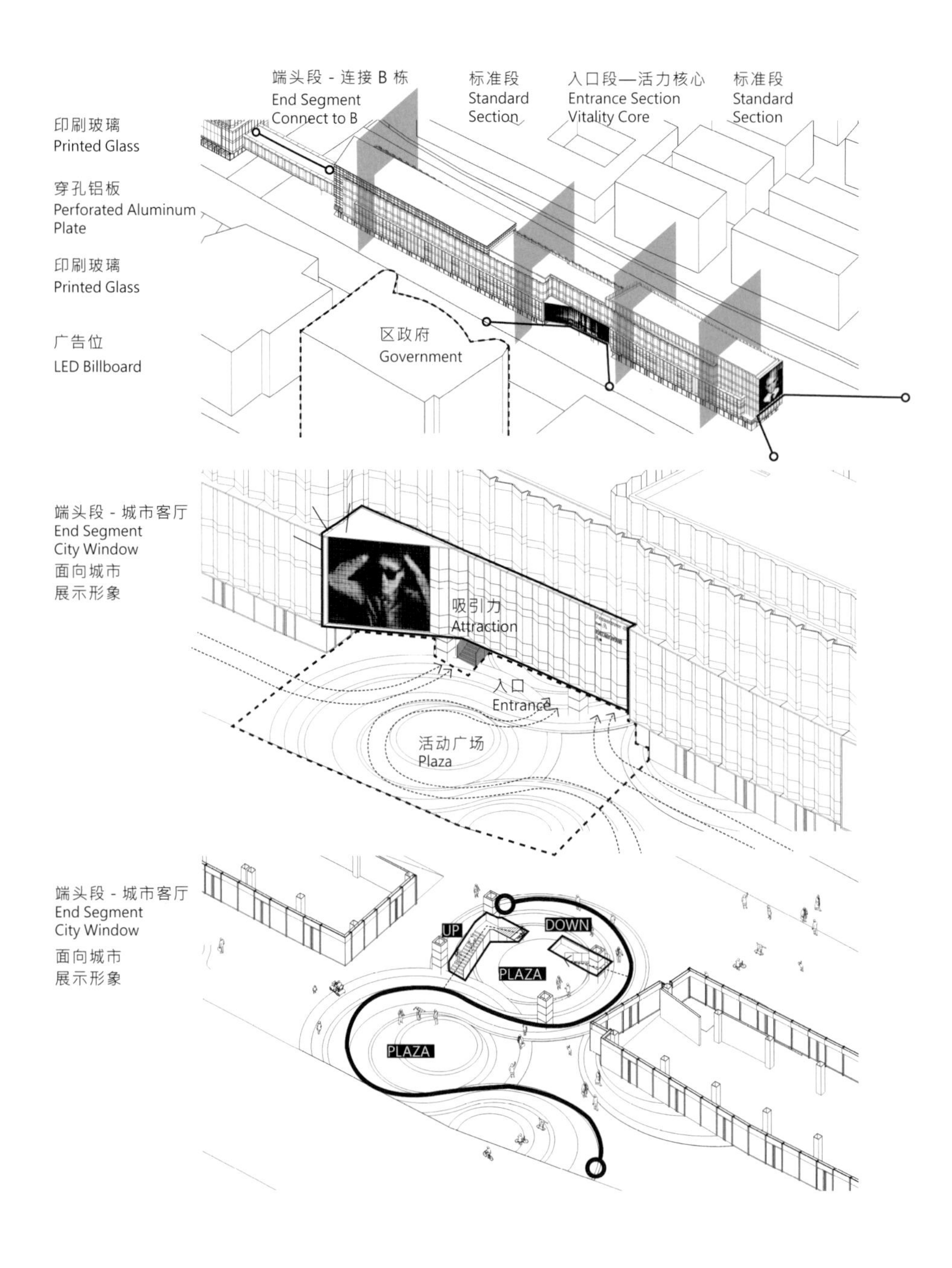
端头段 - 连接 B 栋
End Segment
Connect to B
标准段
Standard
Section
入口段—活力核心
Entrance Section
Vitality Core
标准段
Standard
Section
印刷玻璃
Printed Glass
穿孔铝板
Perforated Aluminum
Plate
印刷玻璃
Printed Glass
广告位
LED Billboard
区政府
Government
端头段 - 城市客厅
End Segment
City Window
面向城市
展示形象
吸引力
Attraction
入口
Entrance
活动广场
Plaza
端头段 - 城市客厅
End Segment
City Window
面向城市
展示形象
UP
DOWN
PLAZA
PLAZA

场地现状
Site Status

场景渲染
Scene Rendering

未来城二期位居重要城市形象节点，西侧面对珠江路与太平北路路口，是整个未来城项目面向城市的形象展示面，中部入口正对玄武区政府是汇聚区域人流的节点，东侧与首开段通过连廊相连接，与首开段形成统一的建筑形象。如何树立重要的空间节点与城市的关系也是我们设计思考的重要切入点。

三处重要的空间节点将未来城将未来城二期在横向分为五段，设计在尊重现有分段的基础上对各个分段进行了重新定义。西端头段，充分利用城市环境，展示项目特点。中部入口，重新梳理空间关系，汇聚活力东部连廊，重新设计连廊及广告展示位，体现未来城的项目的前沿创新感。

The second phase of Future City is located at the important node of the city image. The west side faces the junction of Zhujiang Road and Taiping North Road. It is the extended surface of image that faces the city of the entire project of Future City. The central entrance is directly opposite to the Government of Xuanwu District, which is the node of converging people of the area. The east side and the first opening section are connected by a corridor, forming a unified building image with the first opening section. How to establish the relationship between important spatial nodes and cities is also an important starting point for our thinking of design.

The three important nodes of space divide the second phase of Future City into five segments in the horizontal direction. The design redefines each segment on the basis of respecting the existing segment. At the western end of the term, it makes full use of the urban environment to demonstrate the characteristics of the project. The central entrance reorganizes the relationship of space, gathering vitality in the eastern corridor. The corridors and exhibition spaces of advertising are redesigned to reflect the cutting-edge innovation of the project of Future City.

城市之光
Light of the City

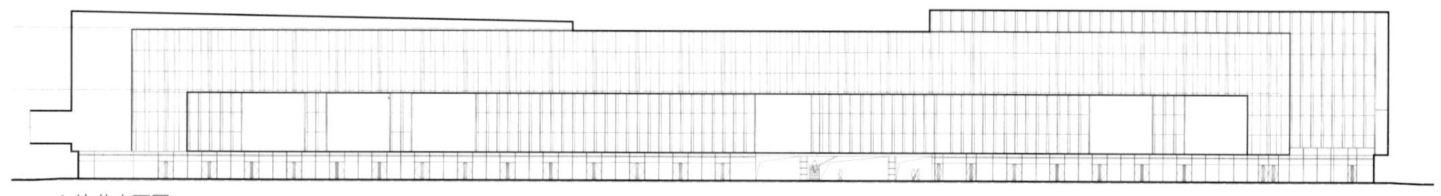

A 栋北立面图
Northern Facade of the A Building

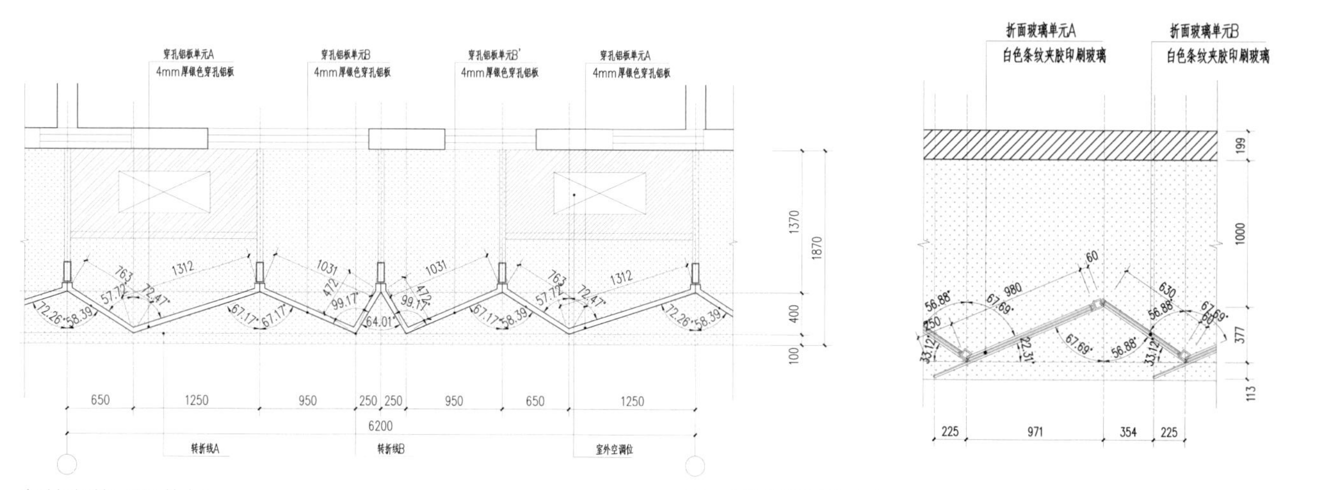

穿孔铝板单元平面放大图
Perforated Aluminum Plate Unit

折面玻璃单元平面放大图
Folding Glass Unit

面向城市环境——城市客厅

未来城二期西端头面向开阔的城市环境，是整个未来城项目最重要的形象展示面，西侧被停车及私建占用的街角广场被重新清理，与端头展示面结合打造出珠江路的城市客厅，把人群汇聚过来形成与建筑空间的互动，提升项目的影响力。

项目的中部入口——活力核心

原建筑入口由于后期使用的不善已经失去了形象昭示性难以汇聚人气，我们重新梳理了入口部分的各项设计元素：立面、广告位、竖向交通体、架空空间、铺地景观、吊顶，将整个空间节点进行一体化设计。最终入口立面简洁大气，通透的折面玻璃幕墙能很好渲染内部商业氛围；底层的架空空间人流与视线不再被阻断，能够沟通建筑南北两面；竖向交通体系明确且具有很好的引导性。经过设计的重新整合，入口部分焕然一新，形成汇聚整个项目人气的活力核心。

Facing the Urban Environment - Urban Living Room

The open environment of the second phase of the west of Future City is the most important exhibition surface of the image of the entire project of Future City. The western corner square that was parked and privately occupied was re-cleared, combined with the exhibition surface of the front end to create living room of the city. The crowd is gathered to form interaction with the building space, enhancing the influence of the project.

Central Entrance of the Project - the Core of Vitality

The entrance of the original building has lost its expression of image due to poor use in later periods and it is difficult to gather popularity. We re-analyzed each element of design of the entrance section: facades, advertising spaces, vertical traffic volumes, overhead spaces, paved landscapes, ceilings, which integrated the design of the entire space node. The final facade of entrance is simple and grand. The transparent folding glass curtain wall can render the internal atmosphere of business well. The flow of people and sight in the underlying overhead space is no longer blocked and can connect both north and south of the building. The vertical system of traffic is clear and well-guided. After the re-integration of the design, the entrance section is completely renewed, forming the core of vitality that gathers the popularity of the entire project.

DAY1 2017,04,20
施工队伍进场 脚手架施工
Construction Approach

DAY8 2017,04,27
一层幕墙龙骨安装
1F Curtain Wall Keel

DAY11 2017,04,30
后置埋件安装
Rear Embedded Parts

DAY14 2017,05,03
折面幕墙龙骨开始安装
Folding Curtain Wall Keel

DAY17 2017,05,06
折面幕墙龙骨安装完成
Folding Curtain Wall Keel

DAY21 2017,05,10
样板段钢结构加固完成
Structure Reinforcement

DAY23 2017,05,12
样板段背衬板施工完成
Backing Plate

DAY24 2017,05,13
开始安装玻璃
Install Glass

DAY26 2017,05,15
启动段大部分折面玻璃完成
Folding Curtain

DAY27 2017,05,16
玻璃盒子超白玻进场
Ultra White Glass Approach

DAY28 2017,05,17
四层折面玻璃开始施工
Top Folded Glass

DAY29 2017,05,18
开始拆除样板段脚手架
Dismantle Scaffold

DAY30 2017,05,19
落架清洗玻璃幕墙
Cleaning Curtain Wall

DAY31 2017,05,20
"江苏省发展大会"呈现
Jiangsu Development Conference

DAY39 2017,05,28
转角钢结构加固完成
Construction Approach

DAY42 2017,05,31
转角玻璃盒子幕墙龙骨安装
Curtain Wall Keel

DAY47 2017,06,05
背衬板后安装保温层
Construction Approach

DAY54 2017,06,12
东立面幕墙龙骨安装
East Curtain Wall Keel

DAY57 2017,06,15
北立面幕墙玻璃安装
North Curtain Wall Glass

DAY59 2017,06,16
街角玻璃盒子背衬板安装
North Curtain Wall Glass

DAY61 2017,06,18
安装街角段玻璃
Corner Glass

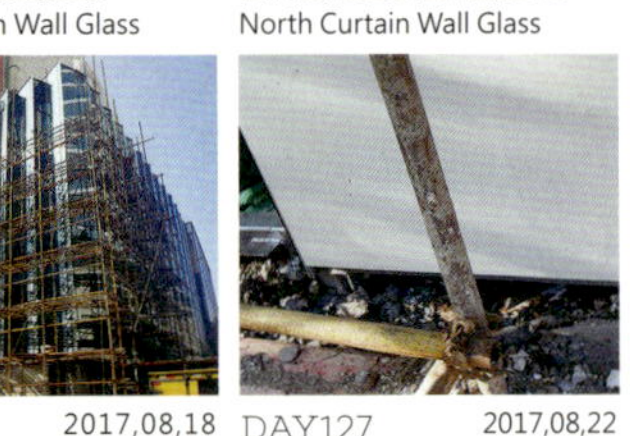

DAY67 2017,06,24
启动区北立面脚手架拆除
North Facade Complete

DAY118 2017,08,13
南立面铝板完成安装
Aluminum Curtain Wall

DAY120 2017,08,15
南立面幕墙玻璃
Curtain Wall Glass

DAY121 2017,08,16
北入口段钢结构加固
Structure Reinforcement

DAY123 2017,08,18
东南角玻璃安装
Curtain Wall Glass

DAY127 2017,08,22
南立面铝板下收口
Curtain Wall Glass

DAY133 2017,08,28
启动区南立面脚手架拆除
South Facade Complete

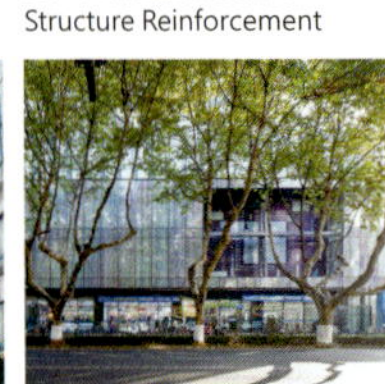

DAY165 2017,09,29
移动段钢结构加固
Structure Reinforcement

DAY177 2017,10,11
彩釉玻璃安装
Curtain Wall Glass

DAY179 2017,10,13
移动段屋顶玻璃安装
Top Folded Glass

DAY188 2017,10,22
移动段脚手架拆除
Facade Complete

DAY192 2017,10,26
东立面一层电箱维护结构
East Electric Box

DAY193 2017,10,27
屋顶压顶施工
The Roof Top Construction

DAY196 2017,10,30
B段主体完工
B Segment Completion

导：
江苏省省委书记

展大会
摩

导：罗康瑞
团董事长

导：穆耕林
区长

区长听取
报会议

投资主体：玄武区珠江路整治管理办公室，南京

Investment Main Body: Xuanwu District Zhujiang Road Renovation Management Office, Nanjing

视察领导：李强　江苏省省委书记 / 张敬华　南京市市委书记 / 徐曙海　玄武区区位书记 / 穆耕林　玄武区区长

Inspection leader:
Li Qiang ,Jiangsu Provincial Party Committee Secretary
Zhang Jinghua ,Nanjing City Party Committee Secretary
Xu Shuhai ,Xuanwu District Party Committee Secretary
Mu Genglin ,Xuanwu District Chief Executive

运营主体：瑞安集团，香港

Operating Subject: Shui on Group, Hong Kong

视察领导：罗康瑞　瑞安集团董事长 / 肖鹰　瑞安建业南京项目总经理

Inspection leader: Luo Kangrui ,Chairman of Shui on Group / Xiao Ying　General Manager of Shui on group Nanjing Project

原设计单位：南京大学建筑规划设计研究院

The Original Design Firm: ADINJU, Nanjing

工程总包：江苏全邦建设工程有限公司

Project Contractor: Quan Bang, Nanjing

幕墙施工：南京金中建幕墙装饰有限公司

Curtain Wall Construction: Jin Zhong Jian,Nanjing

灯光施工：江苏爱涛文化产业有限公司

Lighting Construction: Jiangsu Artall culturaul Industry ,Nanjing

工程监理：南京银佳建设监理有限公司

Project Supervision: Yin Jia,Nanjing

幕墙顾问：弗思特工程咨询有限公司（南京）

Facade Consultant: Forster,Nanjing

结构勘察：江苏省建筑科学研究院检测中心

Structural Investigation: BM LDI, Nanjing

结构顾问：江苏省建筑材料研究设计院

Structural Consultant: BM LDI, Nanjing

照明设计：上海色点照明设计公司

Lighting Design: Locus, Shanghai

室内设计：上海穆氏设计公司

Interior Design: MMoser Associates · Shanghai

方格子华容道 / 兰州城关区两场一馆
Square Klotski / Chengguan District Two Stadium One in Lanzhou

兰州

占地面积　：207 000 平方米

建筑面积　：409 350 平方米

容 积 率　：2.0

设计时间　：2014

Lanzhou

Floor Space : 207,000 m^2

Gross Floor Area : 409,350 m^2

Plot Ratio : 2.0

The Time of Design : 2014

主城区公共配套升级

The Upgrading of Public Facilities in the Main Urban Area

“匠人营国，方九里，旁三门。国中九经九纬，经涂九轨。”

——《周礼 · 考工记》

古人以方格化的网络树立了城市规划的原始模型。兰州这座千年古城，即使不像北京、西安那样严整，但方正守中的中原文化仍深深烙印在文脉中。所以在接手这次 300 余亩中心土地城市设计时，我们首先想到的就是试图还原金城文脉，重塑方正大气的古都形象。

地块位于兰州城关区黄河以北、徐家山森林公园以西，建设总用地面积 20.7 万平方米，省市区三级政府、会展中心、大剧院及体育公园均在其 3 公里辐射圈内。项目包含旅游、休闲、办公、产业、居住、度假六大业态，同时增加风情商业街区、文化体育、医疗、酒店及服务配套功能，打造成兰州的文体综合的区域生活中心。

而将这些丰富的功能融于一张规划网格中，首先就需要轴线明确，分区清晰。地块围绕南北向主轴，分为四大板块——南侧为九年制学校与文体医综合板块，北侧则是文化创意产业园与精品公寓。而后将相应功能置入，划分为尺度适宜的棋盘式布局。南侧为城市服务型公共建筑，以大尺度围合院落再现传统官式殿堂的空间。北侧则更贴近生活化场景，以小尺度多进院落勾画园林式办公园区，以中心围合的大花园规划精品生活住区，以一宽两窄的街巷缝补办公园区与生活住区间的空白。南北地块建筑，业态尺度各异，但仍统一在由中央主轴生成的规划格局中，由功能需求生成不同的密度，并以相似的模度还原古典空间的节奏。而民国风等传统立面手法的渲染，更增强了时代剧般的舞台效果。

网格化是最传统的设计手段。而依附于具体功能的置入，能让暧昧的方格子找到自己合适的空间，这就是城市文脉的张力。正是依附于这种张力，两场一馆才在功能的置入中还原出城市空间原有的节奏。

"Craftsmen camping the country, 4500 meters, next to the three doors. Nine longitudes and nine latitudes are in the country , which passes through nine tracks."

——《Zhou courtesy》

The ancients established a primitive model of urban planning with a grid of networks. Even if Lanzhou is not as rigorous as Beijing and Xi'an, Central Plain Culture is still deeply imprinted in literature. So when we took over urban design of the central land that is more than 300 acres, our first idea is trying to restore Jincheng context and reinvent the image of ancient and grand city.

The plot is located in the north of Yellow River in Chengguan District of Lanzhou and the west of Xujiashan Forest Park. It has a total land area of 207,000 square meters. Three levels of government of the province, city and district, convention and exhibition center, grand theater and sports parks are all within 3 kilometers of radiation circle. The project includes six major formats of tourism, leisure, office, industry, residence, and vacations. Meanwhile, it will increase matched functions of commercial districts, culture and sports, medical services, hotels, and services to create an integrated regional center of living of culture and sports.

When these rich functions are combined in a planning grid, the axis needs clearly defining and the division is clear. The plots revolve around a main axis of north and south and are divided into four sections: The southern plot is the nine-year school and culture, sports and medicine. The northern plot is the cultural and creative industrial park and boutique apartments. We put corresponding functions into it that is divided into appropriate layout of chessboard-type. On the south, there is a service apartment building in the city, which surrounds courtyards on a large scale to reproduce space of traditional official palaces. On the north, it is closer to the scene of daily life. The garden-style office park is drawn down at a small scale. A area of exquisite life is planned with a large garden surrounded by the center. A street with one wide area and two narrow areas is used to make up blank area of office and living. The plots of the north and south have different scales of operation. But they are still unified in the planning pattern generated by the central spindle, generating different densities by requirements of function and restoring rhythm of classical space with the similar model. The rendering of traditional techniques of facades such as the Republican style has enhanced stage effects of the era drama.

Networking is the most traditional method of design. Attached to the placement of specific functions, you can let vague square grid find its own suitable space, which is tension of the context of cities. It is precisely because of the tension that the two halls and one pavilion can be restored the original rhythm of urban space in the placement of functions.

主城区、黄河边、六大公共配套功能
地域传统、棋盘式布局、城市轴线、古坊街区
文体综合区域中心

The Main Urban Area, Near the Yellow River, Six Public and Matched functions, regional tradition, layout of chessboard-style, urban axis, Gufang District
Regional Center of the Synthesis of Cultures and Sports

36°04′
47.60″ N
103°50′
34.78″ E

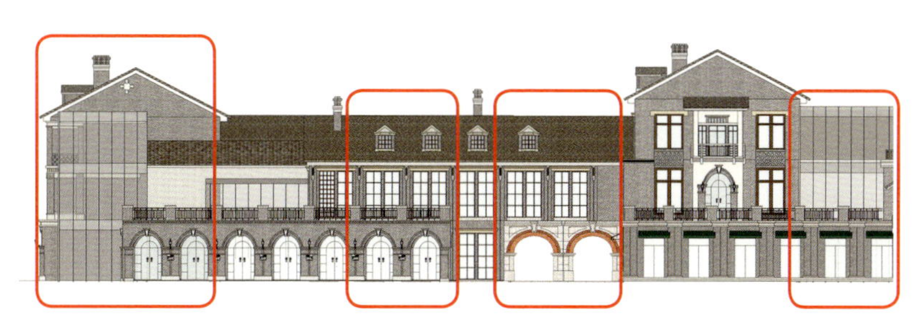

标准段：B-1, B-2
非标准段：F-1,F-2

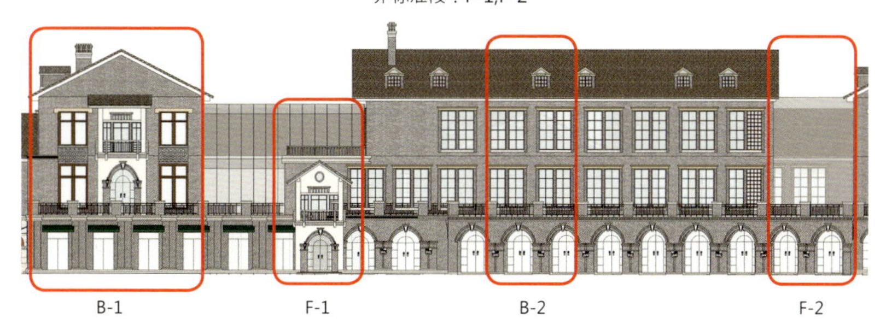

标准段：B-1, B-2, B-3
非标准段：F-1,F-2

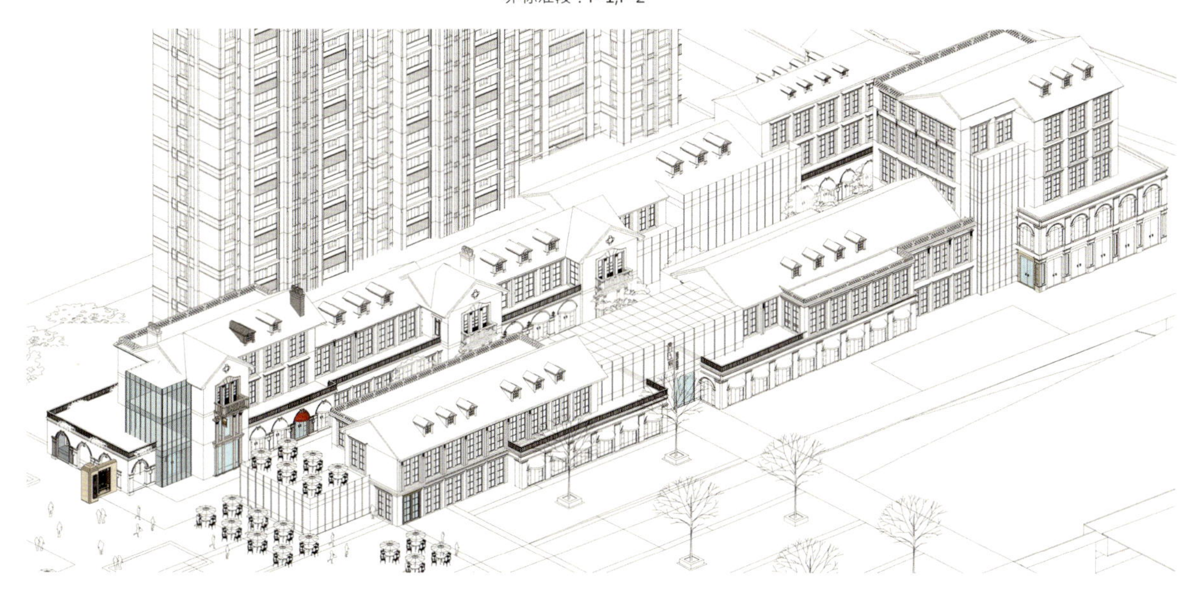

文体综合区域中心：六大主题功能
Regional Center of the Synthesis of Cultures and Sports: Six Thematic Functions

融合地域传统的棋盘式布局方式
The Way of the Chessboard-style Layout That Integrates Tegional Tradition

四大规划理念：

取法自然、古城揽胜——结合古城肌理原型，通过九宫格模数化形成正南北棋盘布局。坊墙里门、宽街窄巷——以主轴为基线，将建筑群组划分为有清晰界限的组团，而街巷作为最活跃的脉络，以张弛自由的形式反映城市肌理。殿堂宇阁、庭院深深——以方正的庭院为母题，将空间大小高低深浅各异的建筑融于棋盘式街巷中。青砖黛瓦、玉池飞虹——以民国风为立面主题，并配以中心水景广场、玻璃挑廊等元素，营造既沉稳又时尚的场所氛围。

Four Concepts of Planning:

We take the natural and ancient city of Range Rover - combined with the texture prototype of ancient cities, through the Jiugong grid module to form the north and south chessboard-style layout. The inside door of the Fang Wall, broad streets and narrow alleys - Using the major axis as the baseline, the architectural group is divided into groups with clear boundaries. The streets and lanes, as the most active context, reflect the urban texture in the form of free strain and relaxation. Palaces and attics, deep courtyard - to the motif of the Founder's courtyard, the space of different levels of height and depth of the building into the chessboard lanes. Blue brick, jade pool and rainbow - the style of the Republic of China for the facade theme, and with the center of the water square, glass gallery and other elements to create a calm and stylish atmosphere of the place.

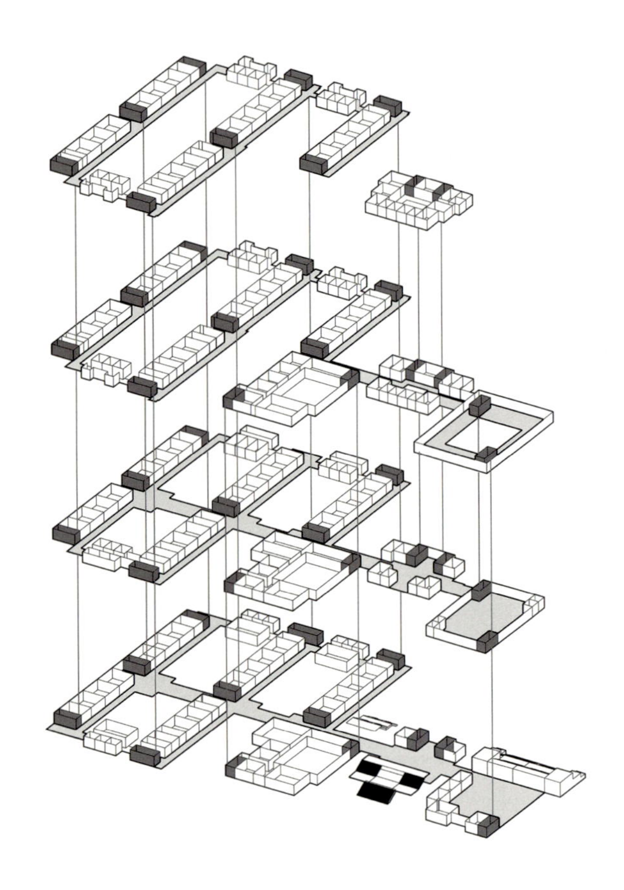

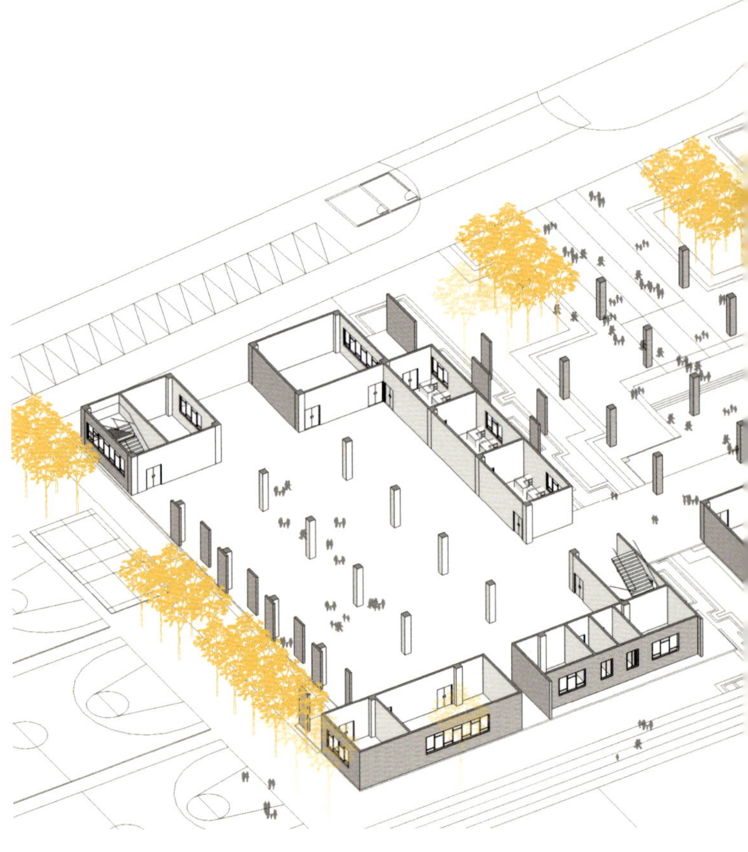

功能流线图
Functional Streamline Graph

低约束的张力 / 上海国际能源创新中心

Less Restrictive Tension / International Energy Innovation Center in Shanghai

上海

占地面积 : 266 400 平方米（400 亩）

建筑面积 : 930 000 平方米

容 积 率 : 3.4

设计时间 : 2015

Shanghai

Floor Space : 266,400 m^2

Gross Floor Area : 930,000 m^2

Plot Ratio : 3.4

The Time of Design : 2015

申能转型再发展

The Transformation and Redevelopment of Shenergy

新时期的机遇

区域开发，永远是深挖现存和过往的因素后，对未来展开的前瞻性思考。上海的土地寸土寸金，封闭式的厂区会阻碍城市多元化的发展，产业与城市的多维融合，将彻底打通发展的脉络。基于环境和土地价值的考虑，宝钢将逐步搬离上海，大场机场也将于“十三五”期间搬迁。宝山区这块土地将被重新规划，在产城融合的基础上接入上海的新活力。申能作为上海的能源界龙头企业，旗下有上海燃气和上海电气等重磅品牌，拥有大量的产业资源，有意将资源导入吴淞煤气厂所占的这块土地，建立一座国际能源中心，以分布式的功能站点和数据中心覆盖方圆 5 公里区域，推动区域升级。

这是宝山的机遇，也是设计切入的契机，工作的重心就集中在探讨政策背景下产业与土地价值的互动升级。我们通过深读政策，梳理现状，借鉴国内外一系列成功案例，逐步勾勒出设计策略。

较低的约束

由于处于探索阶段，项目设计的前期输入条件严重不足。上位控规大量指标空白，用地退界和容积率等开发强度尚缺，甚至申能介入的具体产业项目在设计之初也不明朗。这都需要长时间与业主和专家多方论证，在对土地开发强度、内容置入等方面形成较为清晰的共识后，在可研的基本条件确立后，建筑设计才能正式展开。

同时，这是工业遗迹上实施的战略型改造设计。塑造有浓厚工业气息的特色能源主题产业中心，需要对原有环境进行深层评估，而这些在上位规划中缺乏指导说明。从土地整合到产业集群的设定，我们都需要在模糊的设计边界中，通过反复的背景调研和案例参考，找寻自己的设计边界。

Opportunities of the New Era

The land of Shanghai is highly valued. The closed area of factory will hinder the development of urban diversification. Multi-dimensional integration of industries and cities will completely open the venation of development. Based on consideration of the environment and value of land, Baogang will gradually move out of Shanghai. Dachang Airport will also be relocated during the period of "13th Five-Year Plan". Baoshan District will be re-planned. New vitality of Shanghai will be tapped on the basis of integration of factories and cities. As a leading enterprise in the industry of energy in Shanghai, Shenneng has a number of heavy brands such as Shanghai Gas and Shanghai Electric, having a large amount of industrial resources. It intends to import resources into the land occupied by Wusong Gas Plant and establish an international energy center to cover a 5km-radius area by distributed the site of function and data center, promoting the region upgrading.

This is an opportunity for Baoshan and design entry. The focus of the work is the discuss of interactive upgrading of the value of industry and land under the policy background. Through in-depth policies, we have sorted out current situations and learned from a series of successful cases at home and abroad, gradually outlining strategies.

Low-level Constraint

In the stage of exploration, the input conditions for early design of the project are seriously insufficient. A large number of indicators in upper limit is blank. The strength of development of the retreats of use of land and plot ratio is lack. Even the projects of industry that Shenneng was involved in were not clear at the beginning. This requires long-term arguments with proprietors and experts. After a relatively clear consensus on strength of land development and capacity input, etc., and the established of basic conditions that can be studied, architecture design can be officially launched.

Meanwhile, this is a strategic design of transformation that is implemented on industrial relics. The creation of the industry center of distinctive energy theme with a strong industrial atmosphere requires a deep evaluation of original environment, which is lack of guidance in high-level planning. From land consolidation to the setting of industrial clusters, we all need to find their own boundaries through repeated research of background and reference of cases in ambiguous boundaries.

宝山区潜力发展区域、煤气厂工业遗迹、巨型企业多产业集群联动发展、传统企业遇上城市更新如何破题?

The Area of Potential Development of Baoshan District, Industrial Relics of Gas Plant , Ganged Development of Multi-industry Clusters of Giant Enterprises
How to deal with the problem that traditional enterprises meet the urban renewal?

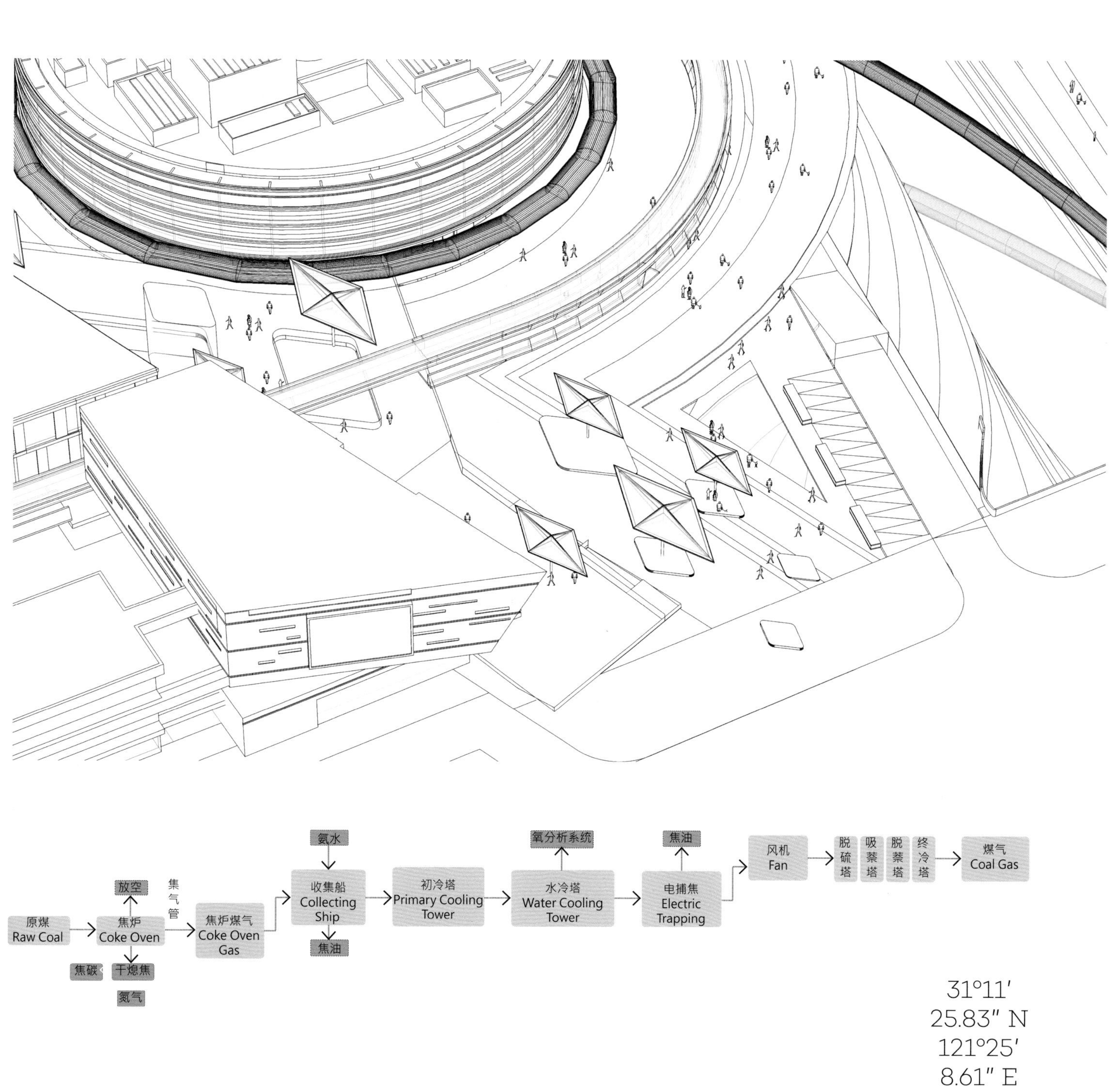

31°11′
25.83″ N
121°25′
8.61″ E

26
19
20
08
14
13
25
10
30
11
29
07
17
32
06
12
09
16
03
18
31
15
04
27
24
02
23
05
01
22
21
生态宜居
Ecological Residence
研发办公
R&D Office
示范交流
Demonstration Communication
平台服务
Platform Service

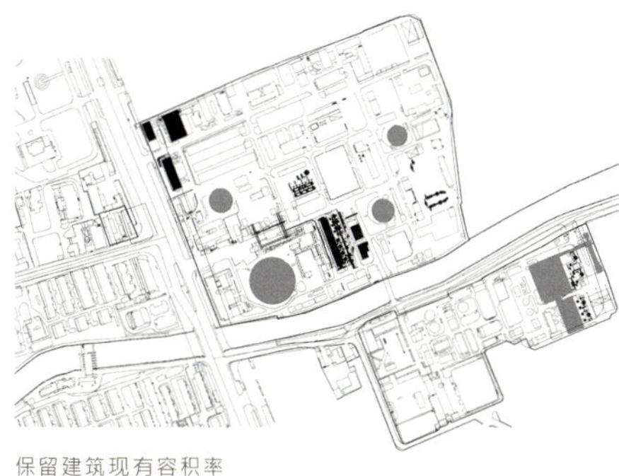

保留建筑现有容积率
Plot Ratio=0.18
Reserve Exsiting Plot Ratio

置入公益空间与核心功能
Plot Ratio=0.40
Placed Public Service Space and Core Function

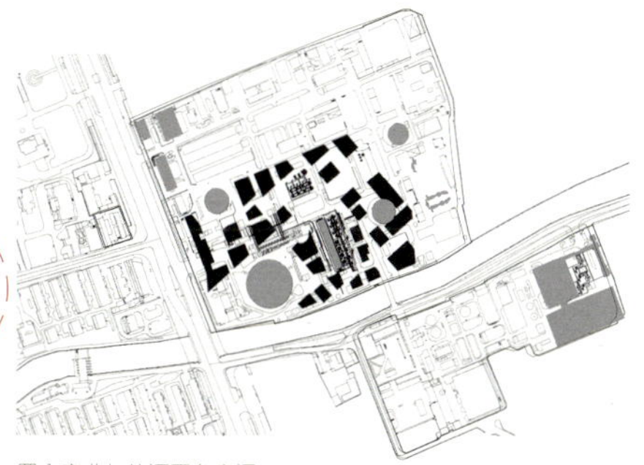

置入商业与能源平台空间
Plot Ratio=0.70
Placed Business and Energy Platform Space

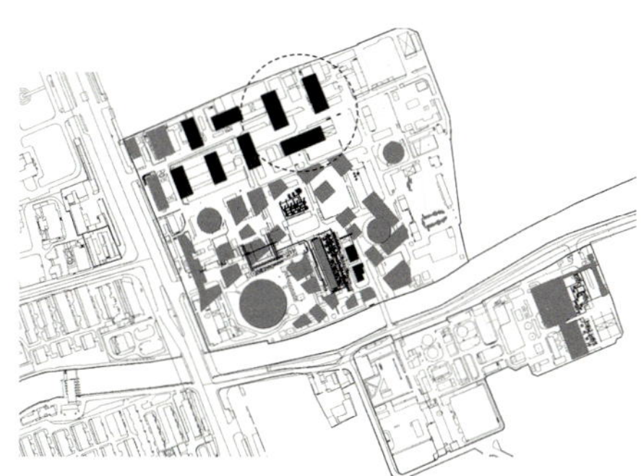

置入人才公寓 + 灵活产品
Plot Ratio=0.99
Placed Talent Apartment and Flexible Product

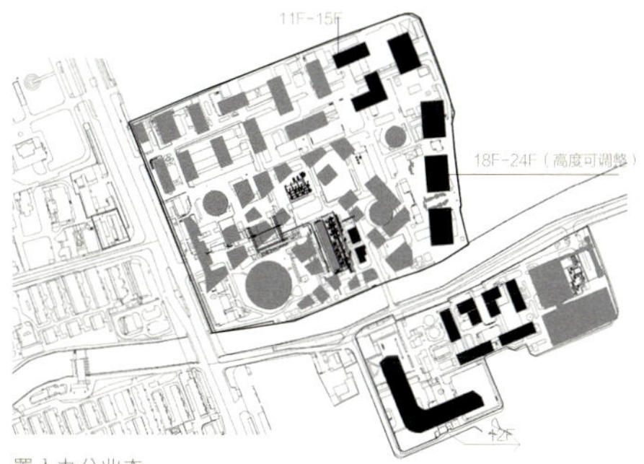

置入办公业态
Plot Ratio=2.46
Placed Office Format

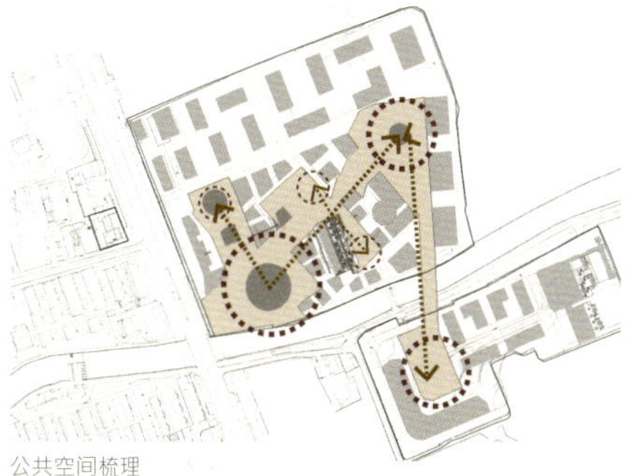

公共空间梳理
Plot Ratio=2.46
Comb public space

01. 未来能源科技展示馆 Energy Innovation Exhibition Hall
02. 行业协会集散中心 Distribution Center
03. 新能源体验馆 Energy Experience Hall
04. 智慧城市示范中心 Demonstration Center
05. 文化展馆 Culture Hall
06. 创新论坛会议中心 Conference Center
07. 创新科技培训中心 Training Center
08. 能源合作组织集散中心 Cooperative Distribution Center
09. 能源科普教育基地 Science Education Base
10. 能源应急产业基地 Industry Base
11. 区域能源供给中心 Energy Supply Center
12. 人才公寓 Talent Apartment
13. 创客空间 Maker Space
14. 能源孵化中心 Incubation Center
15. 能源创新培训中心 Innovation Training Center
16. 能源管理中心 Energy Management Center
17. 能源科技智库中心 Think Tank Center
18. 创新科技工作室 Innovation Studio
19. 能源企业总部 Energy Corporate Headquarter
20. 集成商务办公中心 Business Office Center
21. 健康休闲 Health and Leisure
22. 集成商业服务中心 Business Service Center
23. 能源创新鉴证中心 Innovation Authentication Center
24. 能源交易中心 Energy Trading Center
25. 生态餐饮 Ecological Food
26. 能源研发中心 Energy R&D Center
27. 合同能源管理中心 Contract Management Center
28. 能源综合服务中心 Integrated Service Center
29. 能源数据中心 Energy Data Center
30. 申能智慧云服务中心 Cloud Service Center
31. 保留办公 Reserved Office
32. 工业遗迹 Industry Remains

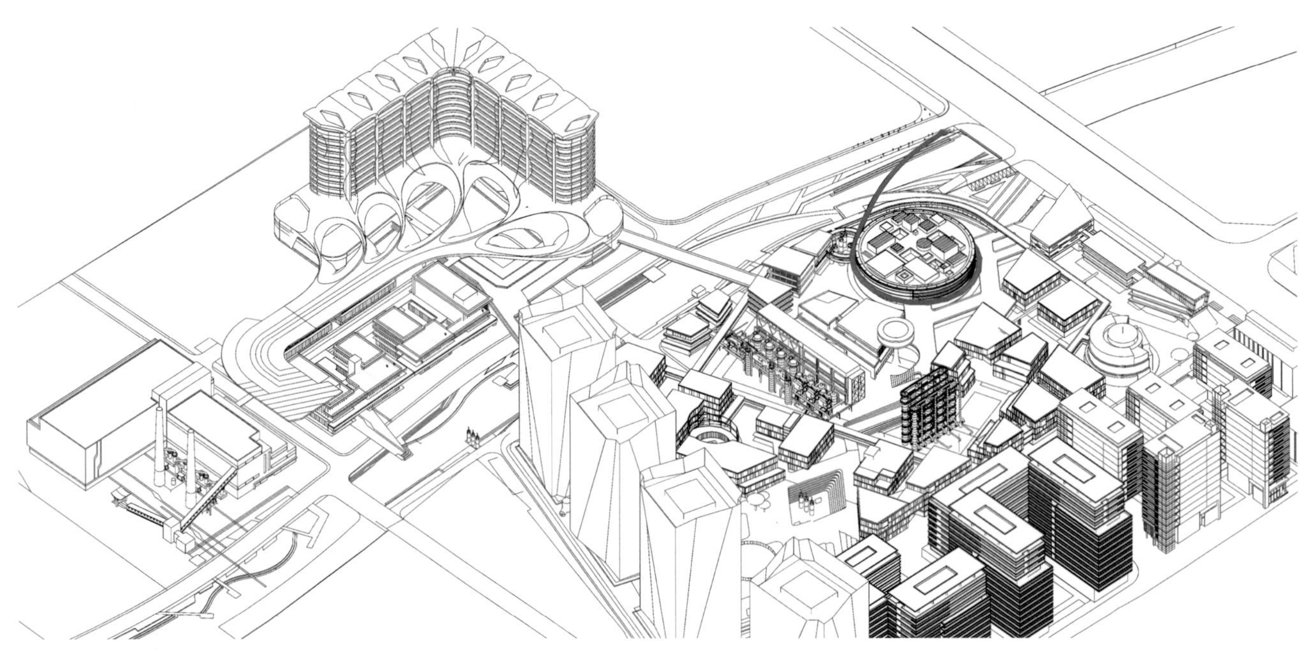

未来能源科技展示馆
The Exhibition Hall of the Energy Technology of Future City

改造依托政策及现存资源

项目位于上海市宝山区的吴淞工业区，由四个分离的地块组成，共400亩。现为吴淞煤气厂的办公、生产及仓储区，废弃建筑较多，且部分地块土地有一定的污染。该区域在宝山区政策引导下正实施整体转型，而吴淞煤气厂位于核心区，如何与政策与申能集团的企业转型战略对接，促进区域功能开发，是重中之重。

存量如何活用 · 增量如何展开？

改造策略

考虑到基地现存的大量工业遗迹，以及现代的开发模式及申能的未来诉求，采取了新建项目结合工业遗迹混合开发的设计策略。

首先，确定保留的历史建筑及工业遗迹，对存量建筑进行分析，测算其容积率。然后，在修缮的保护建筑与设备中置入展示、博览及创新论坛中心等公益性项目，以及其核心功能业态，满足区域公共活动需求，提升片区形象。再通过调研周边市场的产品形态，分析研究形成本项目的功能定位，同时平衡产品类型与容量，高低配多层次开发。最后，通过梳理公共空间，以空间规划策略完善园区环境。

功能业态

项目依托政策支持，结合工业遗迹混合开发，打造以国际能源为主题的产业创新中心示范区；通过产业服务、技术创新及多平台协作打造项目的五大功能板块，实现商旅文“多业联动”，创意创新创业“三创协同”，生产生活生态“三生平衡”，宜商宜游宜业“三宜互促”。

The Related Policy and Existing Resources of Transformation

The project is located in Wusong Industrial Zone, Baoshan District, Shanghai. It consists of four separate parcels that is 400 acres totally. Now they’re office, production and storage of Wusong Gas Factory. There are a lot of abandoned buildings and certain pollution in some plots. The region has implemented an overall transformation under guidance of the policy of Baoshan. Wusong Gas Plant is located in the core area. How to let the policy strategically tie in with the corporate transformation Shenneng and promote development of regional functions is a top priority.

How to Use the Stock? How to Start the Increment?

Considering the large number of existing industy relics at the site, modern model of developing the relics and future demands of Shenneng, the design strategy of newly-built project combined with mixed development of industrial relics was adopted.

First, we determine preserved historical buildings and industrial relics, analyzing existing stock buildings and measuring volume ratios. Projects of public welfare such as exhibitions, reading extensively, and innovation forums centers and core functionalities will be placed in the protected buildings and equipment that is repaired, meeting the needs of regional public activities and enhancing the image of the district. We analyze and research functional orientation through the investigation of product forms of surrounding markets. Meanwhile, we balance the type and capacity, developing it at multiple levels. Finally, we perfect environment of the park with the strategy of spatial planning by tidying the public space.

Functionality

It relies on the support of government and combines the development of industrial heritage to create a demonstration of an industrial innovation center with the theme of international energy: it realize “the linkage of multi-industry” of business, travel and culture through industrial services, technological innovation, and multi-platform cooperation to create five functional blocks. Creation, innovation and entrepreneurship will be integrated. Production, living and ecology will be balanced. Pleasant cities, travelling and business will promote each other.

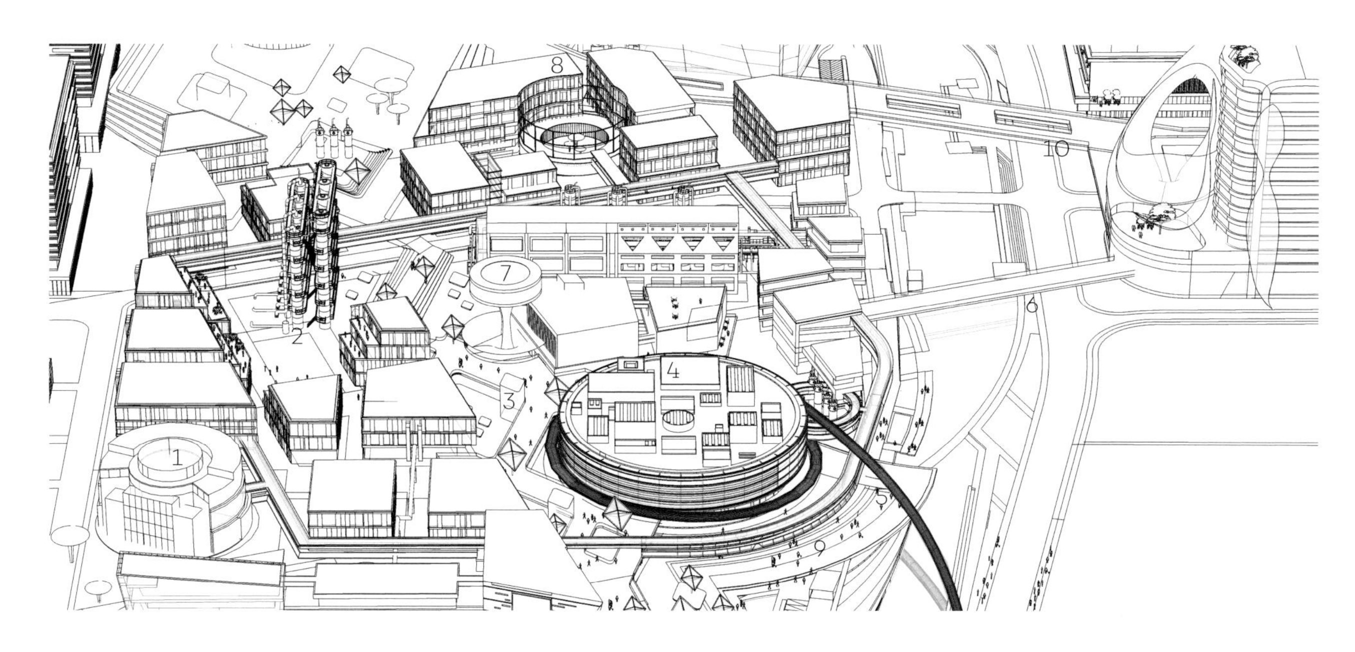

1 能源气球
Energy Balloons

2 脱硫塔
Desulfurization Tower

3 脱硫道
Desulfurization Channel

4 能源中心
Energy Center

5 输煤管入口
Inlet of Coal Conveying Pipe

6 能源一号桥
Energy Bridge No. One

7 水冷塔
Water Cooling Tower

8 原煤中心
Raw Coal Center

9 脱萘环道
Removal of Naphthalene Ring Road

10 能源二号桥
Energy Bridge No. Second

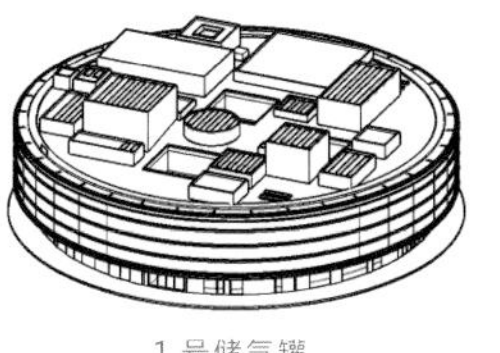
1 号储气罐
Gas Tank No. 1th

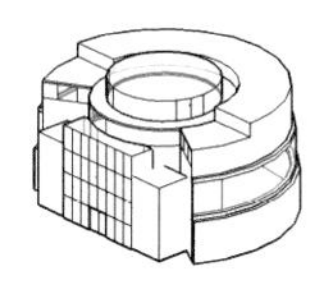
2 号储气罐
Gas Tank No. 2nd

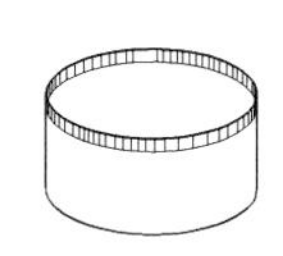
3 号储气罐
Gas Tank No. 3rd

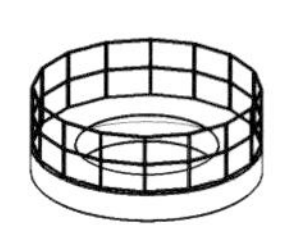
4 号储气罐
Gas Tank No. 4th

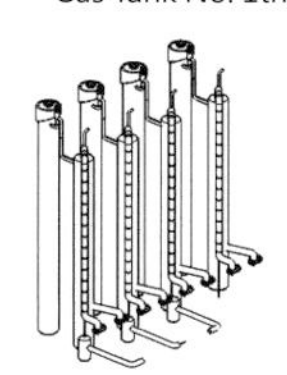
5 号工业遗构
Industrial Relic No. 5th

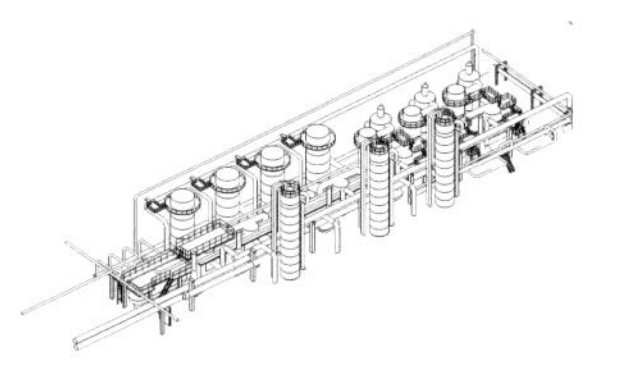
6 号工业遗构
Industrial Relic No. 6th

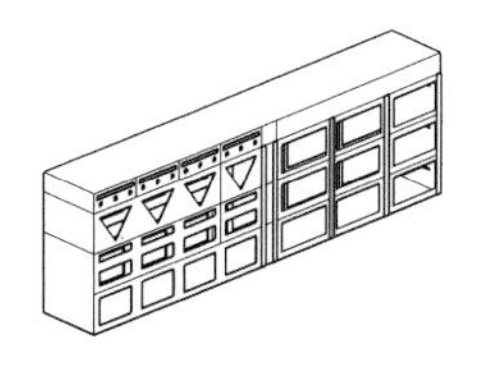
7 号工业遗构
Industrial Relic No. 7th

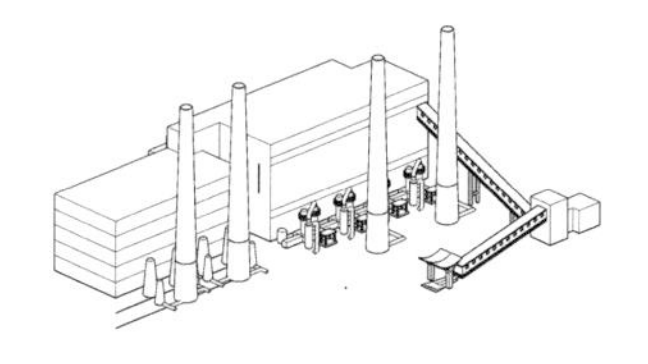
8 号工业遗构
Industrial Relic No. 8th

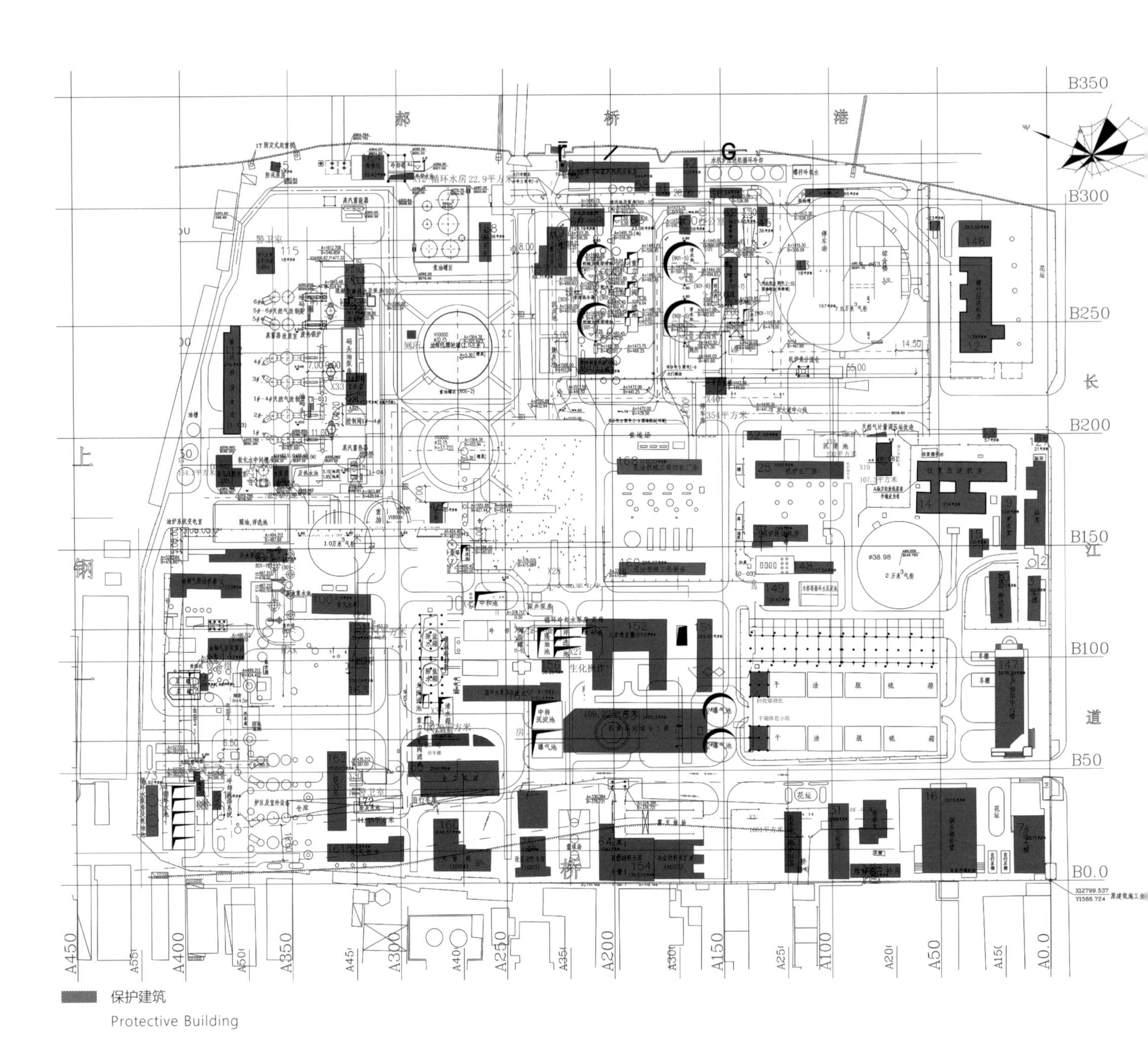

保护建筑
Protective Building

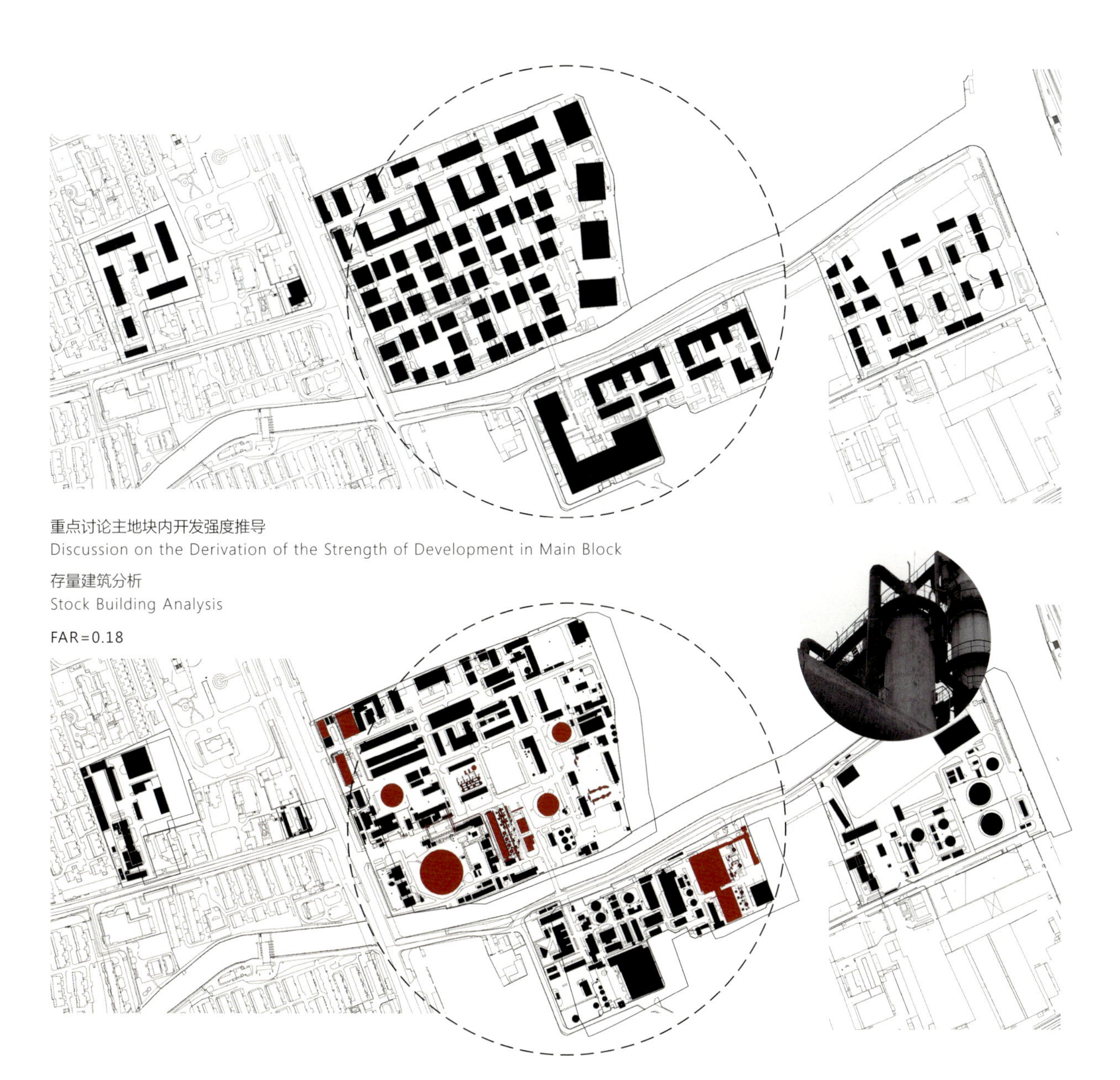

整体开发强度 / 综合五个地块的开发强度，最后综合计算出的开发强度为：容积率 FAR=2.03　总建筑面积：534 000 m^2
容积率 FAR=2.29　总建筑面积：604 000 m^2

Overall development strength/integrated five plots of development strength, the final comprehensive calculation of the development intensity is:
FAR=2.03　Total Floor Area:534,000m^2　FAR=2.29　Total Floor Area:604,000m^2

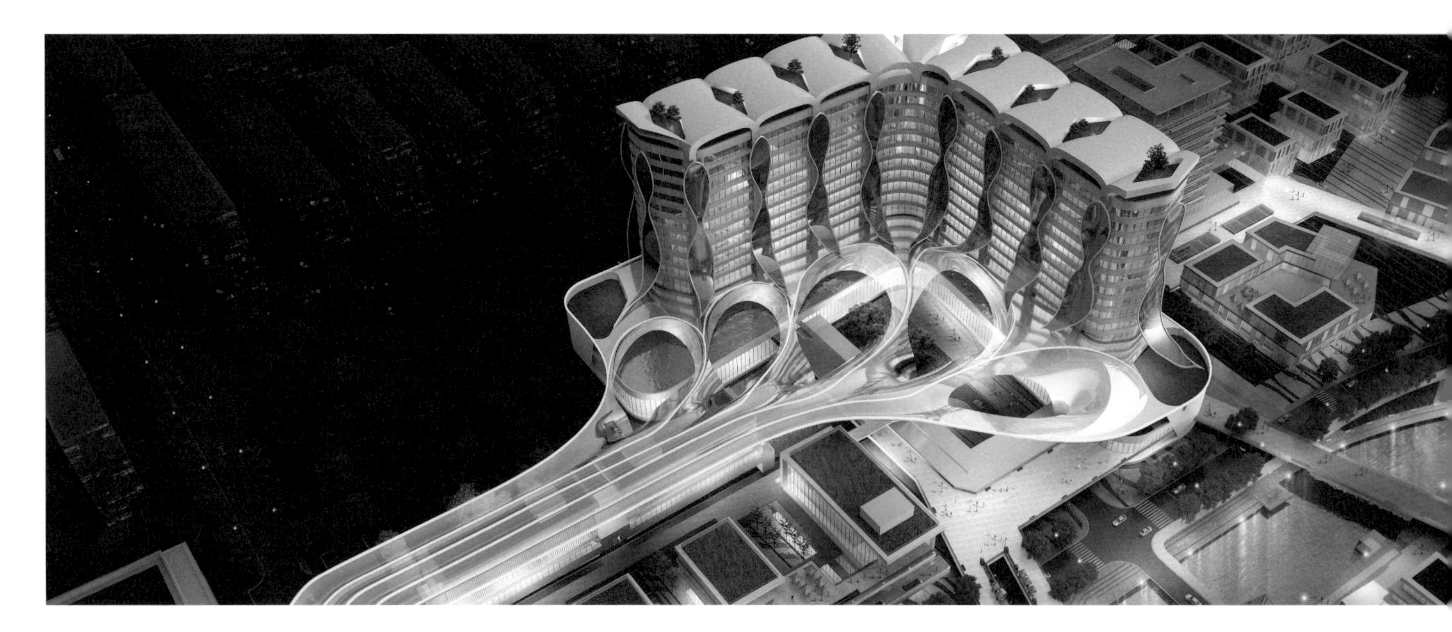

规划策略

基于项目地块相对分散的不利现状，设计通过节点及轴线序列，将各地块紧密联系，打造一条连续的旅游休闲观光流线。在保证开发容量的前提下，项目采用高低配的开发模式，工业遗迹保留的重点区域结合工业遗迹打造花园办公，全拆场地设计高层办公保证项目容积率。

拆留改建

对片区的建筑和工业遗迹现状根据其使用情况和被改造利用的可能性进行评估，将现状遗存分为保留、改建（可以被重新利用）、建议拆除三大类。
保留现状使用情况良好的办公后勤空间、具有突出特征的工业架构；
改建具有良好工业特征且存在被合理改造成有特色的使用空间的工业厂房；
拆除现有遗存中主题特征不明显的工业架构，现存建筑中品质较差或不能被改建成新的使用空间。
确定保留的工业遗迹及历史建筑对总体规划带来一定的影响，如何利用及融入规划设计将成为重点。

景观设计

项目通过景观带划分不同的区域功能，保留具有意义的建筑与设备，形成开放的保护片区。修缮保护建筑与设备并重新植入功能，作为能源展示中心，通过轴线串联，形成空间丰富的工业游览及体验流线。工业遗迹重点保留区域结合多层建筑设计，融合新的改造及建造技术，打造一个低密度的公共艺术办公生态园，升级城市片区功能。

Strategies of Planning

Based on the unfavorable status quo of the project's relatively fragmented land, the blocks are closely linked by the node and axis sequence by the design to create a continuous sightseeing stream of tourism and leisure. Under the premise of ensuring the development of capacity, the project adopts the development mode of high and low allocation. The key areas that are reserved by industrial relics are combined with industrial relics to create garden offices. The high-rise office of the demolition site design ensures the volume ratio of the project.

Demolition and Reconstruction

The status quo of architectures and industrial relics in the area is evaluated based on the situation of using and the possibility of being reconstructed. The existing remains are classified as three categories of reservation, reconstruction (can be reused) and suggestion of demolition.
We retain the office logistics space whose status quo is well-used and industrial structure with outstanding features; We reconstruct industrial plants with good industrial characteristics and being reasonably converted into distinctive spaces of using; We dismantle the existing industrial structures with unobvious characteristics of subject, existing buildings that are of poor quality or cannot be converted into new space for use.
Determining the preserved industrial relics and historical buildings will have a certain impact on the overall planning. How to use and integrate the planning and design will become the focus.

Design of Landscape

The project divides different regional functions through landscape zones, retaining meaningful buildings and equipment, and forming an open zone of protection. We construct buildings and equipment of protection and re-implant functions as an energy exhibition center, connecting through the axis to form a space-rich streamline of industrial tour and experience. Areas of Industrial heritage preservation combine the design of multi-story building, incorporating new technologies of transformation and construction, creating a low-density ecological park of public art office , upgrading the functions of the urban area.

区域供能

项目依托产业和资源优势，结合新能源技术，打造“上海国际能源创新中心示范区”。

设计考虑置入区域分布式能源供给中心和数据中心，形成除宝之云 IDC 及腾讯上海云数据中心外上海第三大数据中心。区域能源供应系统的置入也呼应了申能集团能源产业的历史背景和未来发展的主体定位。考虑到能源供应系统的特殊性，设计将分布式能源供给中心设置于相对独立的主地块东北角的地下，数据中心则位于地上。鉴于分布式能源供给中心夜间利用效率较低，设计考虑通过数据中心 24 小时的持续耗能以保证能源供给中心的综合使用效率。

目前项目已经建成。开盘首日即已售罄。

区域更新的重心是产业，产业升级的内因是市场和政策的驱动，设计从开始就应该有宏观的思维。城市更新的首要课题即土地开发强度的拟定和功能的置入，再通过充分的案例和政策调研作为设计依据。工业遗迹的价值不仅在于储备的土地和特色场所，可持续发展的产业战略更是在这类设计中得到最佳体现。

Regional Energy Supply

The project relies on the advantages of industry and resources, combined with technologies of the new energy, creating "International Energy Innovation Center Demonstration Area of Shanghai."

The design considers the placement of regional distributed energy supply centers and data centers to form the third largest data center in Shanghai with the exception of Baozhi Cloud IDC and Tencent Shanghai Cloud Data Center. The incorporation of regional systems of energy supply also echoes the historical background and the main orientation for future development of the Shenneng Group. Considering the particularity of the system of energy supply, the design places the distributed energy supply center underground in the northeast corner of the relatively independent main site. And the data center is located on the ground. In view of the low utilization efficiency of the distributed energy supply center at night, the design considers continuous energy consumption through the data center for 24 hours to ensure the comprehensive efficiency of utilization of the energy supply center.

The center of regional renewal is the industry. The internal cause of industrial upgrading is the drive of the market and policies. The design should have a macroscopic thinking from the beginning. The first topic of urban renewal is the formulation of the development intensity of land and the introduction of functions. It is then based on adequate cases and policy investigations. The value of industrial relics lies not only in the reserved land and characteristic sites, but also in the sustainable development of the industrial strategy that is best reflected in such a design.

注：

1 吴淞煤气厂于 2014 年 5 月 28 日正式全面停产。根据签订的协议，宝山区将全力支持申能集团转移低端业态和过剩产能，加强与申能集团转型战略的对接，留住申能集团转型新业态。同时，以上海国际能源创新中心项目为抓手，加快吴淞工业区实质性启动和功能性开发。

2 分布式供能是指利用天然气为一次能源，通过冷热电三联供等方式实现能源的梯级利用，综合能源利用效率在 80% 以上，本项目通过结合数据中心的合理功能配置大大增强能源使用效率。数据中心通过分布式能源供给中心的发电余热恰好覆盖所需冷量，仅在尖峰时段需要少量电空调制冷，其集中设置方式可替代一路市电进户，降低三路市电进户的难度，降低对周围高压变电站的需求。

Note:

1. Wusong Gas Factory officially suspended production on May 28, 2014. According to the signed agreement, Baoshan District will fully support Shenneng Group in shifting the low-end business model and surplus production capacity, strengthening the connection with the strategy of transformation of Shenneng Group, retaining its new business of transformation. At the same time, the project of Shanghai International Energy Innovation Center is taken as the starting point. We accelerate the substantial start and functional development of the Wusong Industrial Zone.

2. Distributed energy supply refers to the use of natural gas as a primary energy source, and cascade utilization of cold, heat, electricity and electricity to achieve cascade utilization of energy. The overall energy utilization rate is above 80%. This project greatly enhances the efficiency of energy by combining the reasonable function configuration of the data center. The residual heat generated by the data center through the distributed energy supply center just covers the required amount of cooling. At the peak hours, a small amount of electric air-conditioning refrigeration is needed. The centralized setting method can replace the entrance of one line of city electricity, reducing the difficulty of three-way access to the city and the demand for surrounding high-voltage substations.

讲故事 / 金地未来系研发

Tell A Story / Research and Development of Jindi Future Department

深圳

Shenzhen

疑问在现在 · 策略在将来

人们无法预见未来，但我们所经历的点点滴滴都或多或少与未来在产生联系。这是一个住宅示范区体系的研发。2014年6月3日我们接手了这个项目，6天后，6月9日就要去总部汇报，在接手这项任务时，团队是犹豫的，毕竟，我们住宅设计经验很不足。但是当了解这轮研发的内容主要是概念创新时，我们有自信描摹出符合近年发展趋势的未来系住宅。

未来系属于金地集团六大体系中较新的一个版块，面向刚工作入世的年轻人，属于刚需产品。在深圳，金地已经有项目在操作，希望能将我们的研发成果直接落地，所以也相当重视。我们在极短的时间内展开了充分思考，探索什么是未来的趋势，什么是年轻人社交的状态，怎样激发住区中的生活活力，从服务业态到场景塑造，从建筑形态到构筑小品，在不大的前区广场中营造时尚的社区生活氛围。

设计提出了三个创新的思路，将售楼处置于地下、跨界结合景观构筑及服务管理打造的回家流线、结合电子商务的发展设计相应功能。在空间与网络的双重聚焦中，营造快乐与便捷的生活。

There Is Doubt Now But the Strategy In the Future

People cannot predict the future. But the bits and pieces that we experience are more or less linked to the future. This is the research and development of a system of residential demonstration zone. We accepted the project on June 3, 2014. Six days later, we would go to the headquarters to have a report on June 9. When we accept the task, the team is hesitant. After all, our experience in residential design is inadequate. However, when we understand that this round of research and development is mainly about conceptual innovation, we have confidence in depicting future-oriented residential buildings that meet the trend of development of this year.

The Future Department belongs to the newer section of the six major systems of Gemdale Group, facing young people who have just entered the workforce. They are just needed products. In Shenzhen, Jindi has already had projects in operation and hopes to bring our achievements of R&D directly to the ground. So it also attaches great importance to it. In a very short period of time, we began to fully think about what is the trend of the future, what is the social status of young people, how to stimulate life in residential areas. From the service industry to the setting of the scene, from architectural forms to constructing skits, we create a stylish atmosphere of community living in the small square in the front area.

The design puts forward three innovative ideas. The sales office is positioned in the underground. The cross-border integration of landscape construction and service management is to build the streamline of the home. We design corresponding functions combined with the development of e-commerce. In the dual focus of space and network, we create a happy and convenient life.

住区前瞻性畅想策划
Prospective Imagination Planning of Residential Areas

22°31′
31.21″ N
114°01′
57.50″ E

未来生活、年轻时尚、场景体验、人性关怀
科技生活、新锐造型

Future life, young fashion
Scene experience and human concern
Science and technology life, new shape

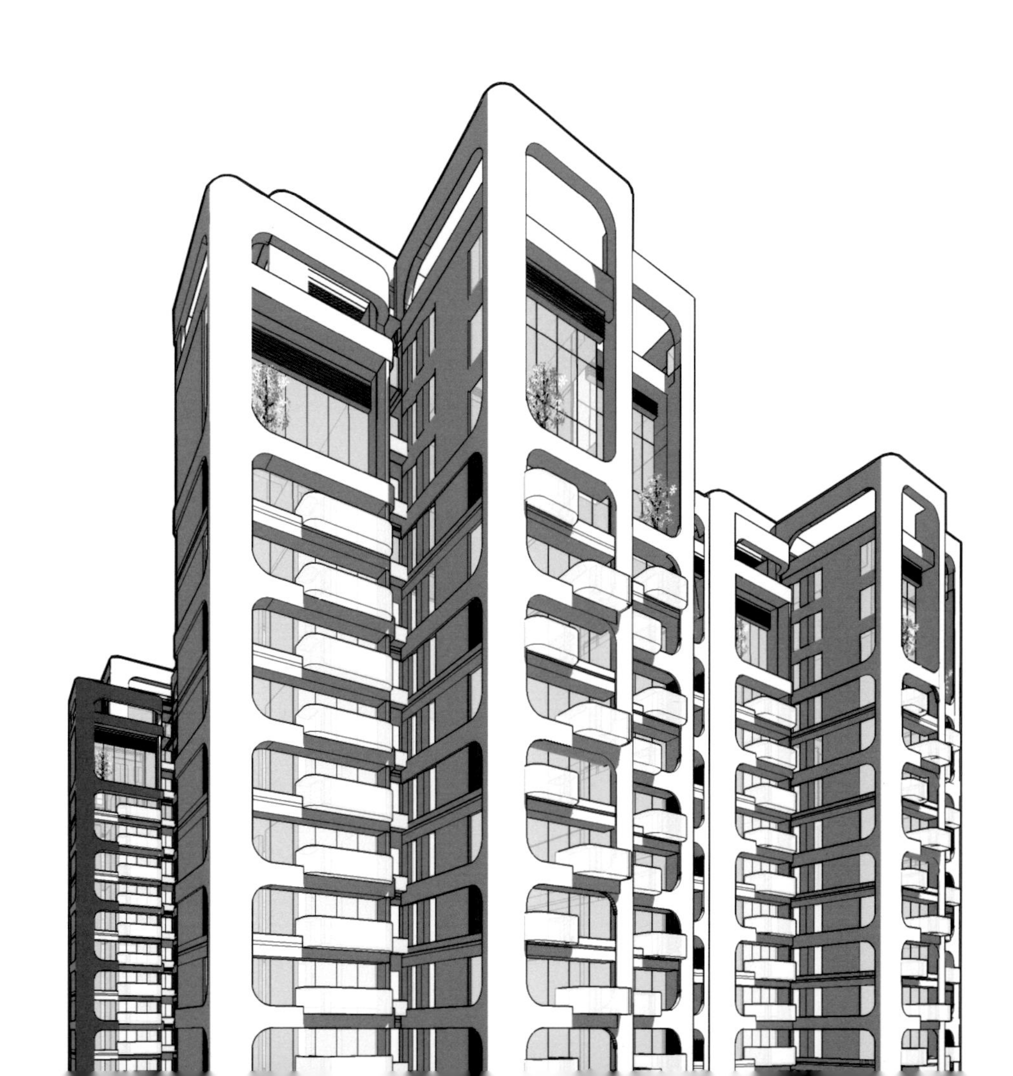

非线性的理性反馈

售楼处下沉，不仅释放了入口广场的空间，也将售楼处外界的立面形象转为室内更浓郁的体验，将启动区的场景营造以较低的成本进行布景化设计，利用陈设将体验推向剧场般的极致效果。

七把以 GRC 材料实现的伞状构筑物，以自由的造型彰显科技感和未来感，在小区门口有序阵列，分别覆盖了下沉会所、共享出行交通站、快递收发、商业及社区综合服务等多项功能。七把伞，不仅是特色入口标识，而且为居民提供了有趣的回家步径，形成了复合性的体验广场。

当时虽是电商发展的早期，但我们坚信未来生活一定会被其深刻影响。所以在小区入口设计了快递收发站点，并相应设计了取送流线及景观。

造型上，我们认为新材料新技术会带来更灵动的设计潮流。设计突破了传统的方形体块，以圆角抹角的 GRC 定制构件穿插体块形成更连续整体的形象，在城市中树立别样的风采。设计或许就应如此，跳出常规的框架，用鲜活的个性与大胆的设想抹平经验冰冷的棱角，勾画不一样的图景。

Nonlinear Feedback of Rationality

The sinking of the sales office not only freed the space of the entrance plaza, but also transformed the external facade image of the sales office into a richer indoor experience, setting up scenes for the start-up area at a lower cost, using furnishings to bring the experience to theater-like and extreme effects.

Seven structures of umbrellas that are realized with GRC materials demonstrate the sense of science and technology and the sense of the future in a free shape. They are arranged in an orderly array at the gate of the community, covering the sinking clubs, shared transportation stations, express delivery, comprehensive services of commerce and community, etc. Multiple functions. Seven umbrellas are not only the hallmarks of special entrances, but also provide residents with interesting homecoming trails, forming a complex square of experience.

Although it was early in the development of e-commerce, we firmly believed that life in the future would be deeply affected by it. Therefore, the courier receiving and dispatching site was designed at the entrance of the community. And the streamlines and landscapes were designed accordingly.

In terms of modeling, we believe that new materials and new technologies will bring more smart trends of design. The design breaks through the traditional square body block, inserts the body block with GRC custom components with rounded corners to form a more continuous overall image, establishing a different style in the city. Perhaps the design should be like this, jumping out of the conventional framework, using the lively personality and bold ideas to smooth out the cold corners of experience, sketching a different picture.

设计模型
Design Model

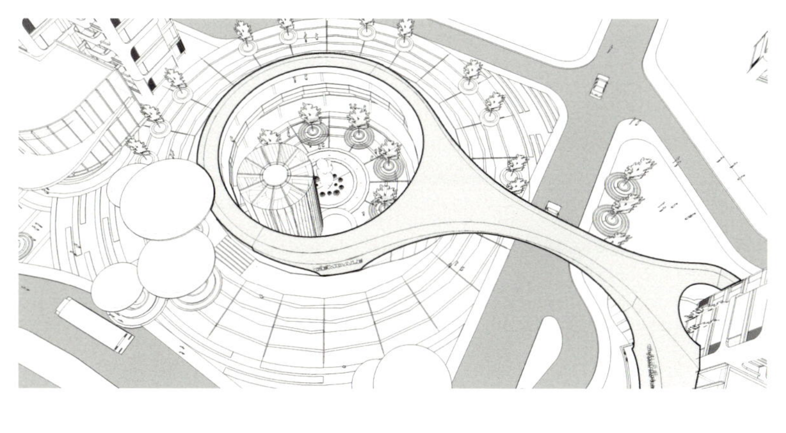

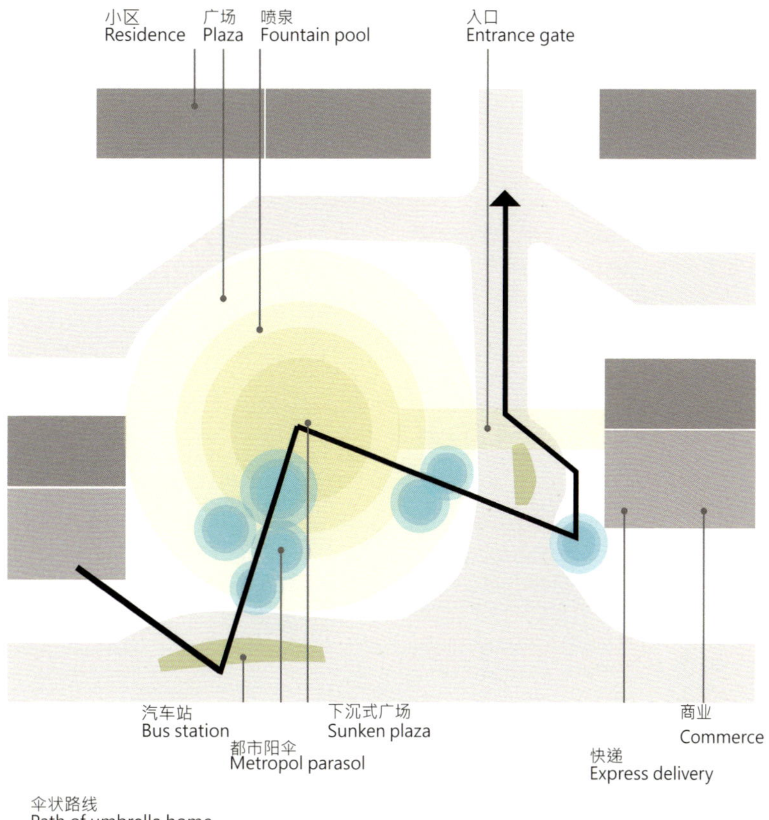
小区
Residence
广场
Plaza
喷泉
Fountain pool
入口
Entrance gate
汽车站
Bus station
都市阳伞
Metropol parasol
下沉式广场
Sunken plaza
商业
Commerce
快递
Express delivery
伞状路线
Path of umbrella home

与树共生 / 长春净水车间改造

Living with Trees / Water Purification Plant Renovation of Changchun

长春

占地面积 ：701 平方米

建筑面积 ：701 平方米

建筑高度 ：6.3 米

设计时间 ：2017

城市丛林的守望者

花径不曾缘客扫，蓬门今始为君开。这是个不足七百平方米的小房子，是长春净水厂的第七净水车间，原址的净水厂建于 1932 年 3 月，当时的伪满洲国刚刚宣告成立，长春被更名为“新京”。为建立、完善这座城市的供水系统，伪国都建设局决定在城市东南处离伊通河一公里的位置修建一座水厂，并命名为“南岭净水场”，即长春净水厂前身。

厂区原有大概 80 多幢建筑，项目用地范围内有 15 幢伪满日期保护建筑，拥有近 27 万平方米生态绿地，其生态价值和文化价值不可小视。该项目也是水石首次尝试的 EPC 项目，新的项目运行方式，新的各团队公司合作模式，也成为此次项目的特殊之处。作为集群设计团队中的一员，成一负责第七车间的更新方案及施工图设计。

七净车间位于厂区的正中心，可谓是藏在城市森林中的树林里的小屋，设计如同在秘境中的私语。整个建筑周边为树木包围，仅有山墙面上的三个角度可视。建筑形体完整而坚固，钢筋混凝土的框架与红砖砌筑的墙体在丛生的绿色中显得极为倔强。走进建筑，内部有 6 个巨大的净水池，池底深达 1.5m，结合原本约 6m 的建筑净高，留下了夹层加建的可能性。我们抚摸着房子和树干，一圈圈环视这个建筑。当时天很蓝，风有些干，太阳扫过我们走过的路径，留下足印与树影的涂鸦。我们意识到，寒冷的冬季中，光与树可以燃起这个设计的温度。

Changchun

Floor Space : 701 m^2

Gross Floor Area : 701 m^2

Building Height : 6.3 m

The Time of Design : 2017

The Guardian of the Heavy Forest of the City

This is a small house of less than seven hundred square meters. It is the seventh water plant of Changchun Water Treatment Plant. The water purification plant on the original site was built in March in 1932. At that time, the pseudo Manchurian state was just announced to be established and Changchun was renamed as "New Beijing". To establish and improve the system of water supply in this city, the Puppet State Construction Bureau decided to build a water plant that was one kilometer southeast of the city from the Yitong River and named it "Nanling Water Purification Plant" that was the predecessor of the Changchun Water Purification Plant.

There are more than 80 existing buildings in the area of factory. There are 15 pseudo-full-date protection buildings within the project site area and nearly 270,000 ecological green spaces. The value of ecology and culture can not be ignored. This project is also the EPC project that W&R has attempted for the first time. The new operation mode of the project and the new cooperation model of each team and company have also become a special part of the project. As a member of the cluster design team, One Studio is responsible for the project of updating and the design of the construction drawing of the seventh workshop.

The Seventh Water Purification Plant is located in the center of the area of factory and is a hut hidden in the woods in the city forest. The design is like a whisper in secret. The entire building is surrounded by trees and only three angles are visible on the gables. The architectural form is complete and sturdy. The frame of reinforced concrete and the red brick masonry wall are extremely stubborn in the bushy green. Entering into the building, there are 6 huge net pools. The depth of the pool reaches 1.5m. Combined with the original building height of about 6m, the possibility of additional mezzanine is remained. We caressed the house and the trunk and looked around the building in a circle. At that time, the sky was blue, the wind was a little dry, and the sun swept through the path we walked through, leaving footprints and shadows of trees. We realized that light and trees can burn the temperature of design in the cold winter.

建造过程
Construction Process

主入口面建造过程，摄于 2017 年 11 月
Main Entrance Surface Construction Process,Taken in 2017,11

北侧建造过程，摄于 2017 年 11 月，修旧如旧的设计
North Side of the Building Process, Repair the Old as the Old Design,Taken in 20

在密林中体味人情的温度

与树共生，与其说是建筑设计的要求，不如说是这个联合设计中所有设计师共同的心声。我们也不例外。树，不仅有木头的质感，光影灵动的斑驳，疏密有致的轮廓，还有家与归来的标志。一个院子，几棵树，窸窸窣窣的风声，袅袅炊烟下家人的招呼，这些就是中国最传统的家的图像，也是最具东方韵味的温暖。而其中最具特质的就是树影下打开的门扇。

在山墙面的主入口，门扇打开间隙透射的光，从密林中探出，我们希望能再现这样的场景。散开的门扇在白天映下长短不一的影子，在晚上延伸出角度各异的光柱，如同一根羽毛轻轻地叠在地上，让整个建筑更加轻盈。在纵向上，我们将六个水池联系起来，并向外扩展，形成内外融通的下沉庭院，在光的映衬下整个建筑如同漂浮起来。在内部空间，我们设置的夹层让空间多了折叠与变化，增强整个设计的趣味性。

这里曾是一座守望长春的水厂。如今，在蔓延的绿色中，我们希望这里仍能蓄存一份安谧的温度。

Appreciating the Temperature of Human Feelings in the Jungle

Coexisting with the tree is not so much a requirement of architectural design as it is the shared voice of all designers in this joint design. There is no exception among us. The tree not only has the texture of wood, the mottle of light and shadow, the outline of density, but also the symbol of home and returning. One yard, a few trees, the sound of buzzing winds and the greetings of family members under the smoke are these images of traditional Chinese homes and the warmth of the most oriental charm. The most characteristic of these is the open door under the shadow of trees.

At the main entrance to the gable, the transmitted light emerges from the jungle when the door is opened. We hope to reproduce this scene. The diffused door leaves a shadow of varying lengths during the day and extends a light beam of different angles at night that, which is like that a feather gently stacked on the ground, making the entire building lighter. In the longitudinal direction, we connect the six pools and expand outwards to form a sinking courtyard with internal and external communion. Under the light, the entire building is like floating. In the interior space, the mezzanine that we set up allows space to be folded and cultured to enhance the fun of the entire design.

It used to be a water plant for Changchun. Today, in the spreading green, we hope that we can still store an ampoule of temperature here.

现场照片，树荫环绕与进入路径
Live Photo, Shade Surround and Access Path

设计模型
Design Model

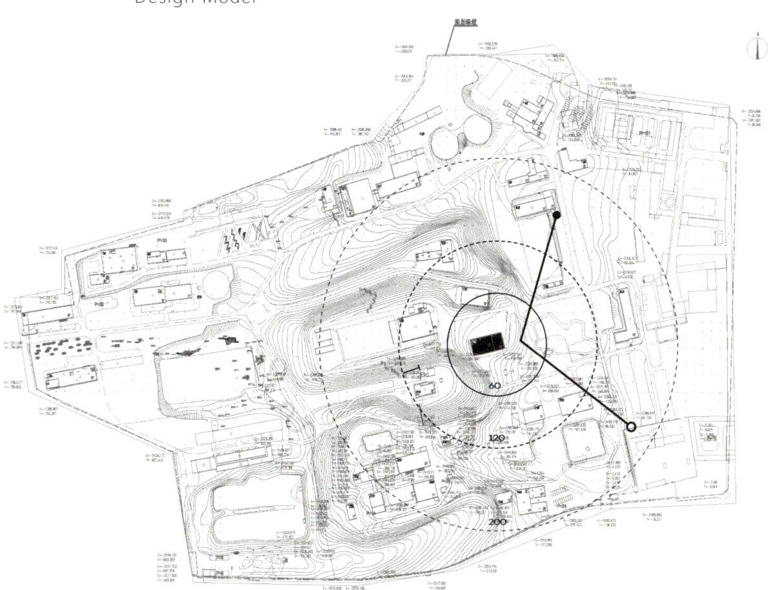

地上地下 / 青岛金地世家启动区

Above and Below Ground / The Promoter Region of Jindi Qingdao Family

青岛

占地面积 ： 3 713 平方米

建筑面积 ： 2 361 平方米

建筑高度 ： 14.6 米

时间 ： 2017

Qingdao

Floor Space : 3,713 m^2

Gross Floor Area : 2,361 m^2

Building Height : 14.6 m

The Time of Design : 2017

1. 虚：玻璃盒子
Virtual:Glass Box

虚：玻璃条窗
2.Virtual:Glass Bar Window

虚：玻璃竖窗
3.Virtual: Glass Vertical Window

实：石材片墙
4.Solid: Stone Wall

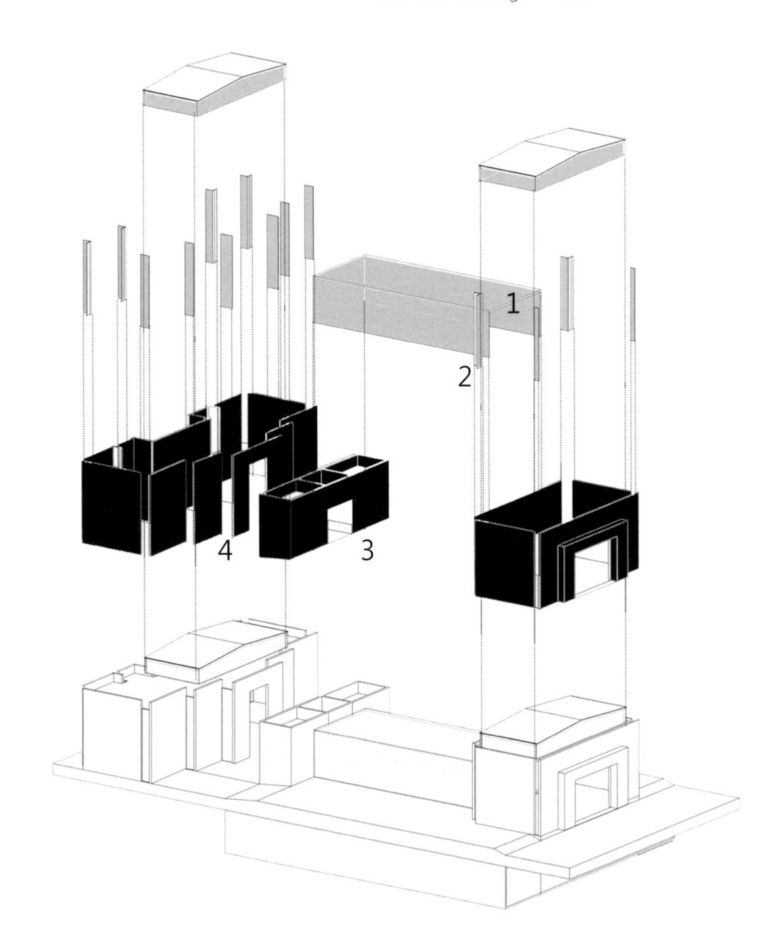

古典比例的现代营造
The Modern Construction of Classical Proportions

36°06′
21.99″ N
120°27′
41.73″ E

城市青岛、一线海景、岛居
金地高阶产品系、紧约束尺度背景下的地下空间利用
基于古典的微度造型创新、短周期

Qingdao, first-line seascape, island house
The underground space utilization in the background of high order products and compact constraints、Based on classical micro - scale modeling innovation, short period

世家：金地旗下最高阶产品系列。

若没有足以传世的建筑，庄园或许仅是一片土地而已无论哪个时代，庄园的建筑都承载着满足世家阶层身份需求、审美需求、生活需求和传世需求的责任。

项目位于黄海之滨的青岛，作为金地世家系列重点战略布局项目，本项目的启动区决定了整个项目气质与调性。天光会所与礼序节奏轴线等创新设计，将时代的生命力注满每个空间和细节。此刻的建筑，不再仅仅是房子，更是世家阶层审美和生活追求的完美表达。

建筑立面并没有完全遵循传统的法式庄园风格而是赋予了其更多的时代特性。提取传统的法式立面元素，运用更为现代的设计元素，更为简洁的手法将其重新设计，通过材质与氛围精到使用烘托世家系列所承载的美学需求。现代与传统结合的设计风格赋予了世家系列新时代的生命力。

Family: The Series of Products in the Highest Stage of Jindi

If there is no building that can be handed down, the manor may be only a piece of land. No matter what era, the architecture of the manor carries the responsibility of satisfying the needs of status, aesthetics, living, and inheritance of the family class.

The project is located in Qingdao that is on the bank of the Yellow River. As the key strategic layout of series of Jinji Family, the initiating zone of the project determines the temperament and tonality of the entire project. The innovative design of the Skylight Club and the rhythm axis of etiquette and order will fill every space and detail with the vitality of the times. The building at the moment is no longer just a house, it is a perfect expression of the aesthetics and the pursuit of life of the family class.

The facade of the building does not fully follow the traditional style of French manor but gives it more characteristics of the times. Extracting elements of traditional French facades, using more elements of modern design, and redesigning them in a more concise manner, through the use of to use The aesthetic needs that is held up by the series of family are presented by the appropriate use of materials and ambience. The combination of modern and traditional styles has given the family a new era of vitality.

工作渲染草图：入口空间设计
Rendering Sketch : Entrance Space Design

三分比例，十分恢宏

项目的设计灵感来自法国的古典建筑。建筑设计充分汲取法国宫廷建筑的精粹，对建筑品质精心打造，在延续法式建筑鲜明、高昂、华丽、风雅的设计底蕴的同时，依靠品质、空间、文化、价值、细节等元素构筑和营建世家这一高端品牌，为业主塑造一个尊贵的顶级社区以及体现身份与价值的平台。

建筑师延启动区中轴规划了礼序景观—大堂—天光会所—售楼处多重空间体验，随着空间的递进人也能感受到空间的收放节奏。最终整个序列在售楼处前广场达到高潮，进入售楼处黄海尽收眼底。建筑外立面采用法式古典“三段式”设计，体现庄重、典雅的风格，顶部以出挑的檐口，线脚勾勒出简洁、明快的建筑体型。为了体现出建筑的当代性，舍弃了大量传统法式建筑的装饰构件，通过石材与玻璃的虚实对比以及材质本身的表现力来烘托整个建筑的贵族气质。与传统法式的梦莎式屋顶不同，项目中使用了深色双坡印花玻璃屋顶，在遵循“三段式”的同时避免了传统法式屋顶沉重的观感，使得整个建筑更加的轻盈现代。

Three-point Ratio, Ten-point Excellence

The design of the project was inspired by the classical architecture of France. The design of architectures fully captures the essence of architectures of French palace and is carefully crafted with respect to architectural quality. While continuing the distinctive, high, gorgeous and elegant design heritage of French architecture, we construct and build the high-end brand Family based on quality, space, culture, value, and details, creating a distinguished top-level community and a platform that shows identity and value for proprietors.

The architect plan multiple spatial experiences of the ritual landscape - Lobby - Skylight Club - sales office along the axis of the initiating zone. With the progressive people in space, they can also feel the rhythm of reunion of the space. In the end, the entire sequence hit a climax at the square in front of the sales office and entered the sales office in a panoramic view of the Yellow Sea. The facade of the building adopts a French classical design of "three-stage", which embodies a dignified and elegant style with a top-of-the-line sculpting and a simple, crisp construction outline. In order to reflect the contemporary nature of the building, a large number of decorative elements of traditional French architecture were abandoned. The aristocratic temperament of the entire building was magnified through the contrast between the actuality of the stone and the glass and the expressiveness of the material itself. Unlike the traditional French-style dreamshade roof, the project uses a dark double-slope printed glass roof, which avoids the heavy perception of traditional French roofs while adhering to the "three-stage" style, making the entire building more light and modern.

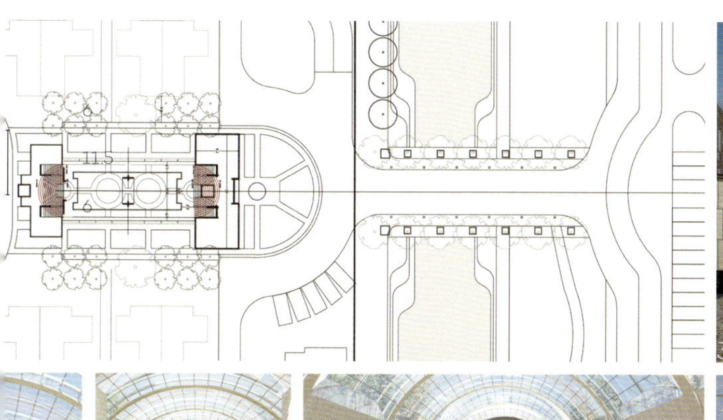

图尺度关系
e Scale Relation of Total Graph

2 室内空间渲染
Interior Space Rendering

3 局部场景
Partial Scenes

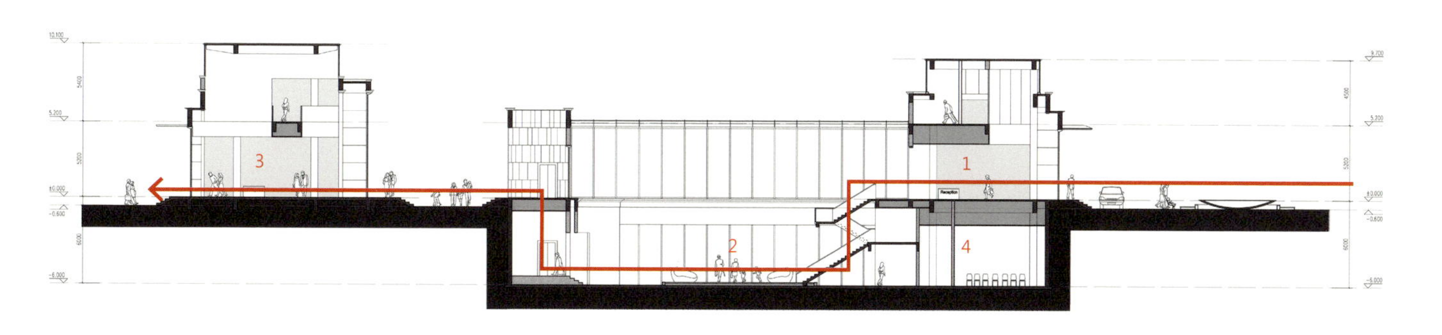

1. 会所
Club

2. 下沉中庭
Sinking Atrium

3. 售楼处
Sales Office

4. 多功能厅
Function Room

锚入风土
Anchoring into the Customs

40℃的水，是热水还是凉水？这取决于之前的体温；一顿饭，好吃还是难吃？这取决于现在饥饿的程度。所有感受都是相对的，而在相对中作为参考的就是风土。

风土，就是一种以公认默许的姿态呈现于人们面前的语言。它不仅是建筑设计的自然背景、文脉联系，也是设计面对尚不成熟的市场不得不与之共舞的规则。创意为了在现实中生根，首先就需要在纷繁的设计环境中找寻扎根的地平线。

然而设计的生根，并不只是简单置入风土，而是将创意的天际线锚入这根地平。毕竟，无意念的建筑只是房子而已。设计可以通过改变用户心中的参照物，深刻影响价值的评价体系。就像古玩店门口放块镇店之宝，让客户在赞叹不已后再进店，即使购买一个钥匙扣都会赢得巨大的价值感，大部分产品线中的旗舰产品，也是为了彰显品牌的价值观而刻意打造。

我们尊重风土，因为这是这个市场这个当下大家公认的交流方言。然而，我们拒绝简单遵从，因为我们想在这片汪洋中抛下一个锚，告诉所有合作伙伴设计的价值。

Galerija
Gallery
ギャラリー

观九洞 观肆
"Nine holes" Scenery with Hills and Waters

张尙媛
Yiyuan，Zhang

镜片，麻纸，水墨、碳
Lenses，Hemp Paper, Ink, Carbon
原尺寸 Original Size 33mm × 33mm
2017

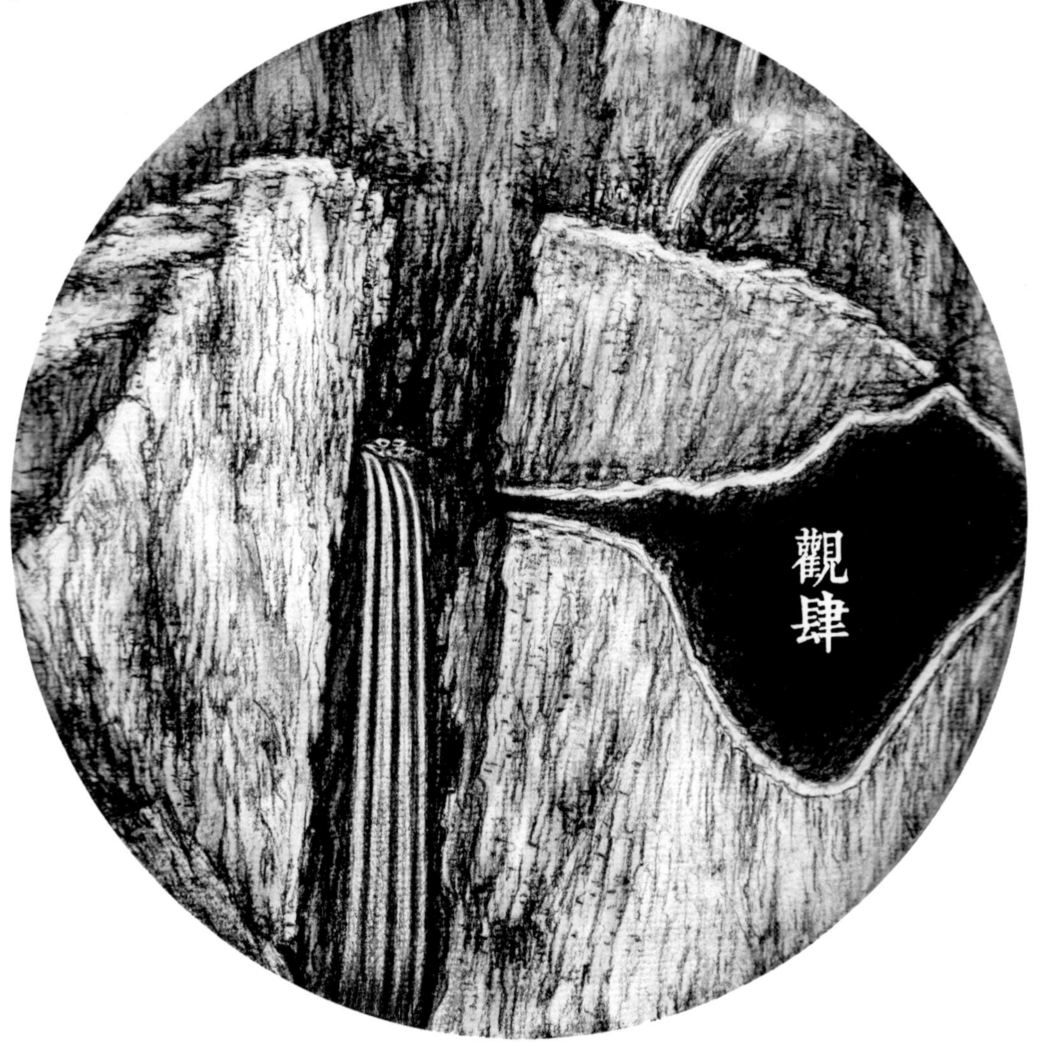

再造园林 / 苏州金鸡湖 Indigo 精品酒店（五星）

Rebuilding Gardens/ Jinji Lake Indigo Boutique Hotel in Suzhou (Five-Star)

苏州

占地面积　: 21 445 平方米

建筑面积　: 57 159 平方米

建筑高度　: 40 米

设计时间　: 2015

Suzhou

Floor Space : 21,445 m^2

Gross Floor Area : 57,159 m^2

Building Height : 40 m

The Time of Design : 2015

传承东方园林文化精髓

Inheritance of the Essence of Oriental Garden Culture

“园林巧于因借，精在体宜，愈非匠所可为，亦非主人所能自主义；需求得人，当要节用。”

——《园冶》

这座坐落在苏州金鸡湖畔的五星级酒店，自出生伊始就注定与文化与城市深深结缘。苏州，仅仅简单的两个字已经包含了很多期待，最有内涵的千年东方古城，最新锐的江苏经济重镇，以最广的胸怀包容现代发展与文化历史。而金鸡湖作为中国最大的湖泊公园，既是苏州最美的景色之一，也是苏州产城融合特色发展的明鉴。在这样的文化、自然、城市格局的坐标中，这座酒店的诞生早已超越平凡，它需要更鲜明的展示这座城市的律动、这方水土的文脉、这个时代的品质。

世界顶级酒店管理公司英迪格为其运营，苏州两千五百年文化为其添色，金鸡湖万亩风景为其喝彩。在周边已忘却奢华的环境中，这座酒店的诞生注定与众不同。

"The garden is clever in borrowing, refined in fitness, the more you can not be a craftsman, nor the owner can be self-reliant: when you need someone, you need to use it."

- 《Yuanye》

This five-star hotel, located on the banks of Suzhou Jinji Lake, was destined to become deeply involved with culture and the city since its inception. Suzhou, the simple two-character words contain many expectations. It is the most connotative city of millennium of the East, the most up-to-date economic town of Jiangsu, embracing the modern development and cultural history with the broadest mind. The Jinji Lake, as the largest lake park in China, is one of the most beautiful scenery in Suzhou. It is also a clear reflection of the development of the characteristics of the city of Suzhou. In the coordinates of such a culture, nature and city pattern, the birth of this hotel has already surpassed the ordinary. It requires a clearer demonstration of the rhythm of the city, the context of this land and water, and the quality of this era.

Indigo, the world's top company of hotel management, has operated for it. The culture of Suzhou that has been 2,500 years adds luster into it. The landscape of Jinji Lake acclaims for it. In an environment where the surrounding luxury has been forgotten, the birth of the hotel is doomed to be different.

苏州、金鸡湖畔、国际一线特色酒店品牌
业主情怀与品质诉求、短时间内如何诠释文脉
品味文化与湖景资源的最佳融合与创新？

Suzhou, on the bank of Jinji Lake, the brand of international first-line hotel
Proprietors' appeal of feelings and quality, how to interpret context in a short time
The best fusion and innovation of tasteful culture and resources of lakescape

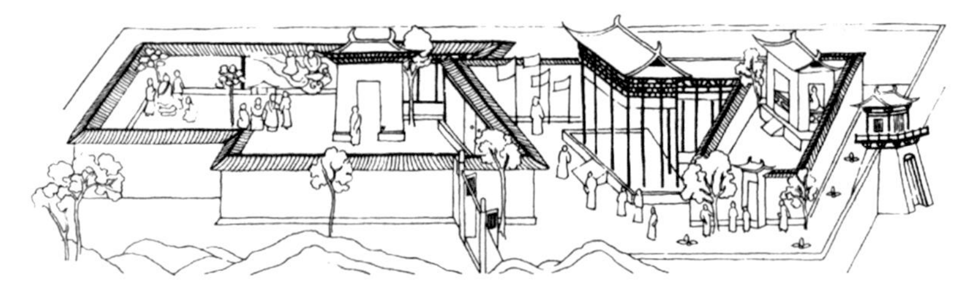

传统园林影壁

Traditional Garden Screen Wall

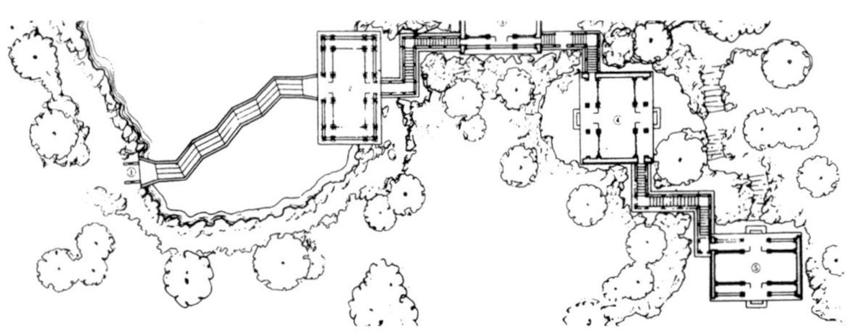

传统园林造景

Traditional Landscape Architecture

31°18′
7.87″ N
120°42′
2.97″ E

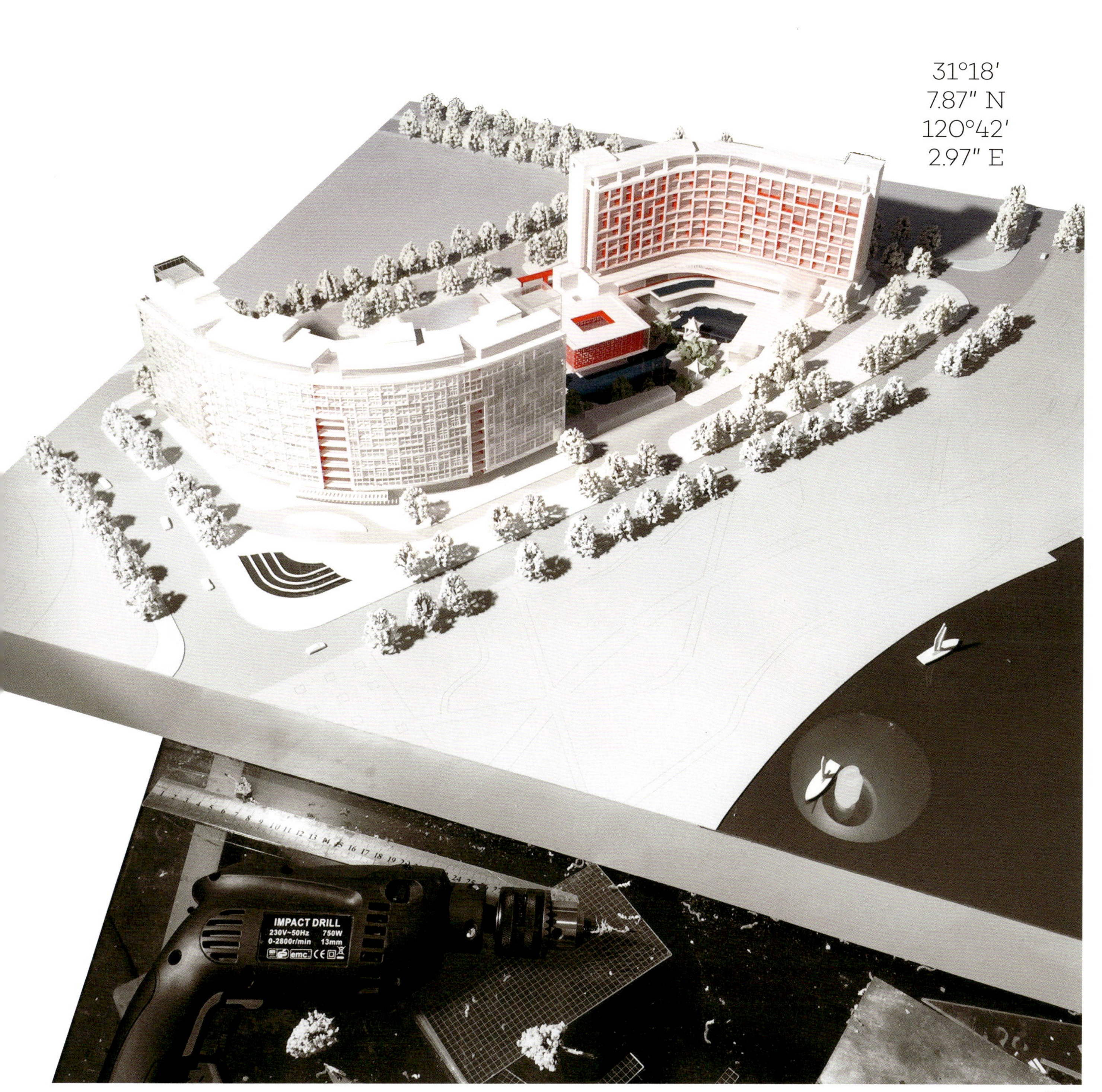

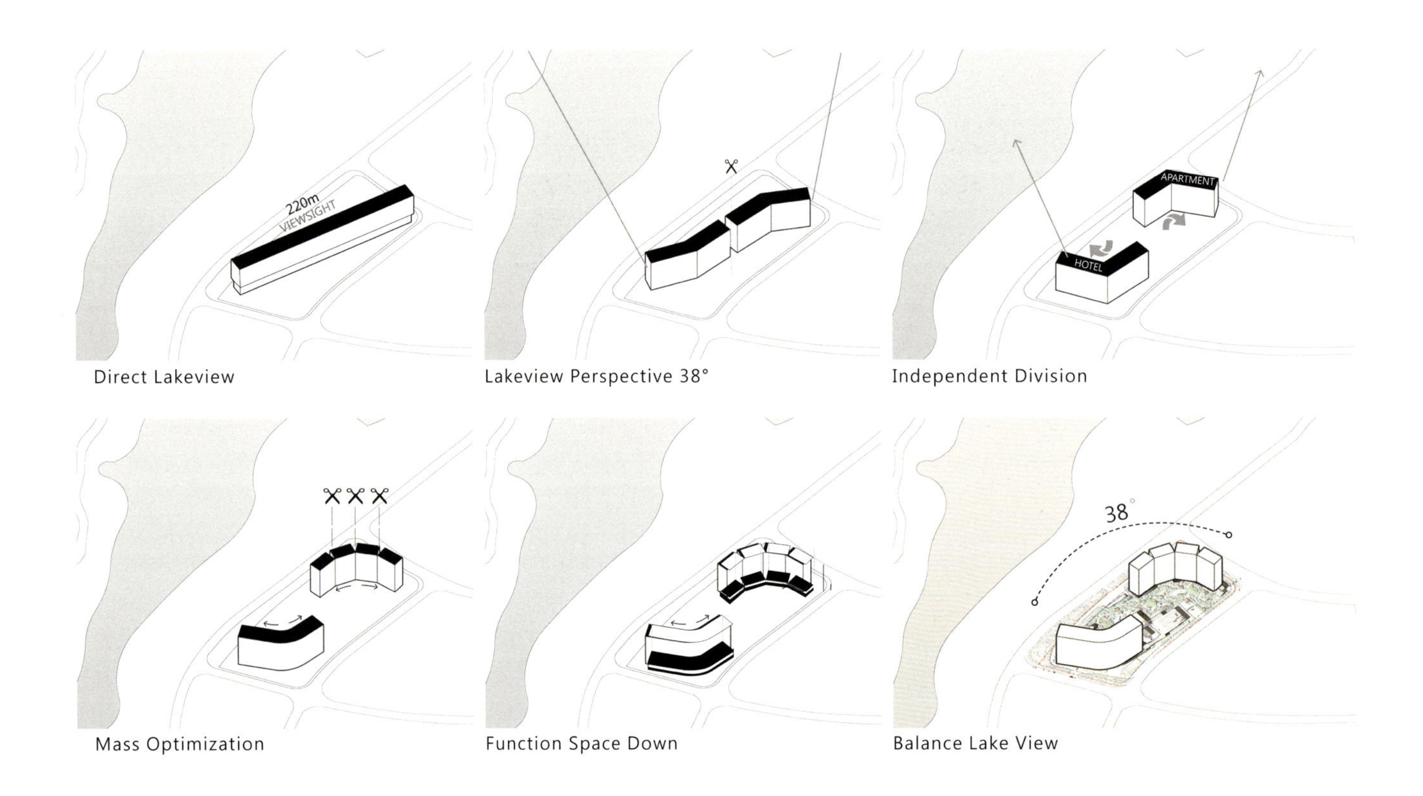

Direct Lakeview

Lakeview Perspective 38°

Independent Division

Mass Optimization

Function Space Down

Balance Lake View

如何凸显与苏州和金鸡湖的关系？

苏杭自古就是人文旅游的重要站点，而现代苏州最重要的景观就是金鸡湖。项目基地就在紧邻金鸡湖畔不足百米距离处，观湖视角达 90° ，最佳观湖视角有 38° 。金鸡湖湖景自然就是最核心的资源。

但是，该地块略低于路面，且与金鸡湖之间的滨湖绿地上有大量树木，高度达 12 米，意味着酒店四五层之上才享有优秀的湖景。加之有 40m 的建筑限高要求，高价值的建筑空间更为稀缺。在这种资源极度稀缺的前提下，我们提出了两大策略，一是低层一定要有景观、景观体验多重化的，二是高度等于价值、极大利用价值面。

最大化利用景观，首先体现出的就是总图形体的生成策略。在经历直面湖景、弯折延展湖景面、酒店公寓独立分区和优化形态的四大步骤后，我们确立了最终体块关系——公寓沿路布置直接面湖，酒店则充分利用内景丰富体验。客房上置，底部为裙房，将视野在立体维度上分级，内庭与外景互动，使最大化的景观价值得以升华。一座尽显东方神韵的园林内庭，无边无际尽收眼底的金鸡湖景，在内外高低的共鸣中诠释了这座酒店最完美的苏州印象。

How to Highlight the Relationship Between Suzhou and Jinji Lake?

Suzhou and Hangzhou have been an important site for cultural tourism since ancient times. The most important landscape of modern Suzhou is Jinji Lake. The project base is located at a distance of less than 100 meters from the shore of Jinji Lake. The viewing angle of the lake is 90 degrees. The best viewing angle is 38 degrees. The landscape of Jinji Lake is naturally the most essential resource.

However, the site is slightly lower than the road surface. And there are a lot of trees on the lakeside green land between Jinji Lake. The height is 12m. These mean that the hotel enjoys excellent lake views on the 4th and 5th floor. In addition to the requirement of the limited height of 40m, high-value building space is even more scarce. Under the premise of extremely scarce resources, we have proposed two major strategies. First, there must be multiple landscape and experiences of landscape at the lower levels. Second, the height is equal to value and the aspect of value is greatly utilized.

To maximize the use of landscapes, first of all, it is the generation strategy of the total graphics. After undergoing the four steps of facing the lake, bending and extending the view of lake, independent partitioning of the hotel apartment, and optimizing the form, we established the relationship of final body-mass - the apartment is directly laid out along the road. The hotel makes full use of the rich experience of the interior. The upper part is the lobby and the bottom is the podium. The horizon is graded on the three dimensions. The inner court interacts with the scenery to maximize the value of the landscape. A landscaped courtyard full of Oriental charm and the landscape of Jinji Lake that has endless panoramic view interpret the most perfect impression of Suzhou of the hotel in the high, low, inside and outside resonance.

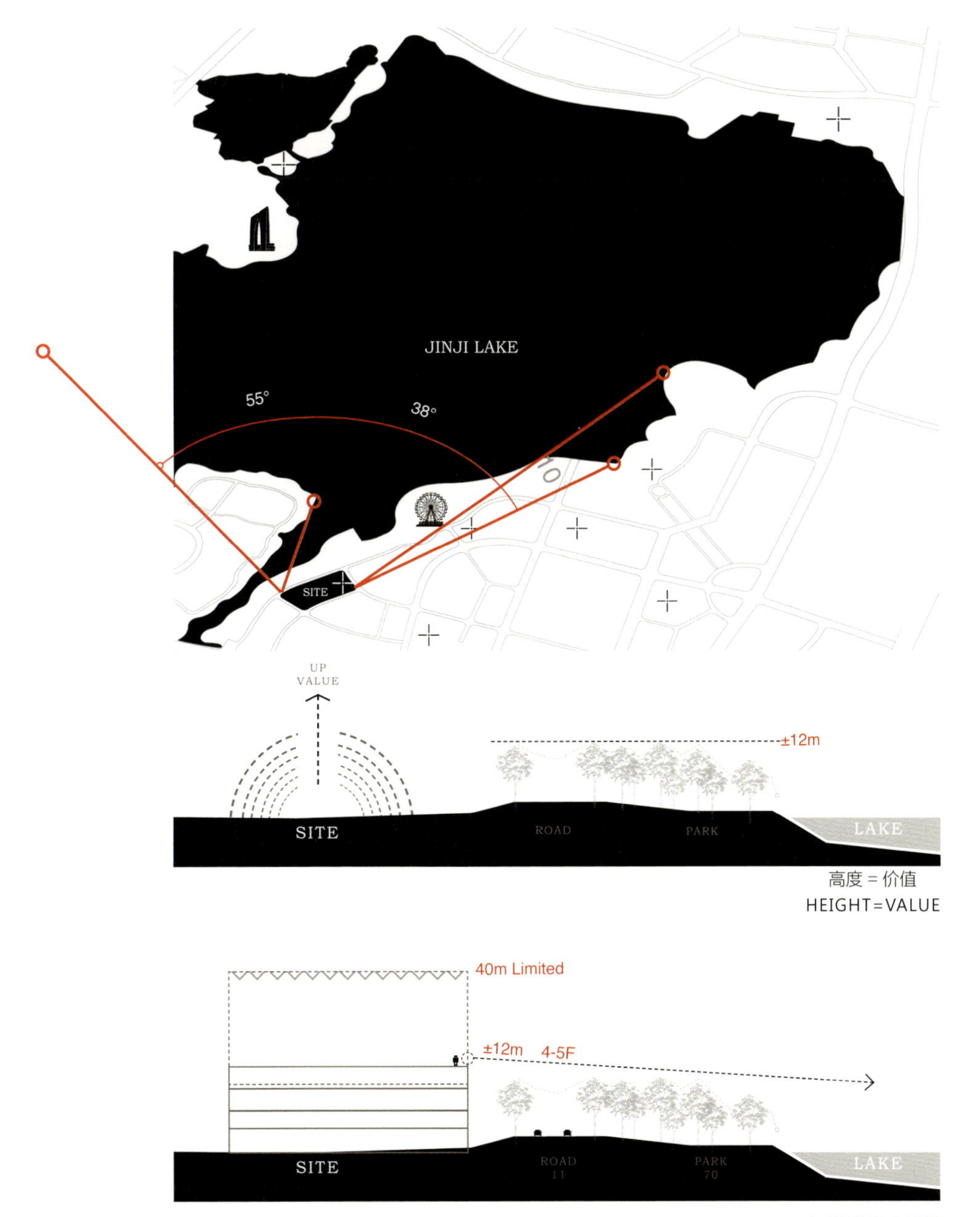

高度 = 价值
HEIGHT=VALUE

(<5F)低层也要景!
SIGHTVIEW ANALYSIS

榭，凭借景而设；景，可穿可贯，可卧可转

Pavilions, by Virtue of the Scene; Scene, Can Wear Penetration, Can Lie Can Turn

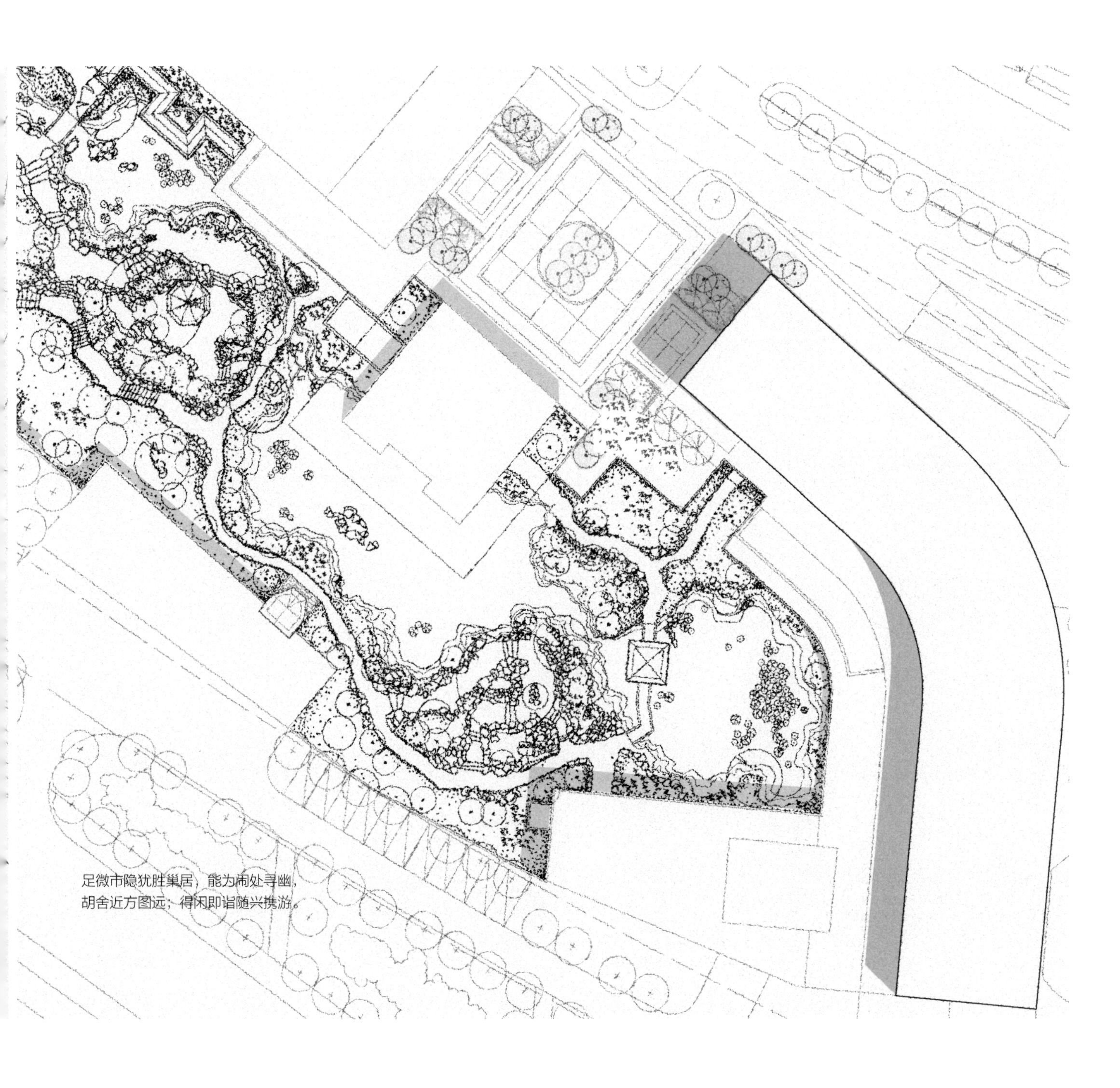

足微市隐犹胜巢居，能为闹处寻幽，
胡舍近方图远；得闲即诣随兴携游。

夜色下的酒店客房与私享园林
Hotel Rooms and Private Gardens at Night

金鸡湖湖心岛观地块“对视”视野
View From Eyot of Jinji Lake , All Rooms Visible

酒店园林一角鸟瞰
Aerial View of One Corner of the Hotel Garden

酒店博古架式窗花立面
The Antique Shelf Style Windows Facade

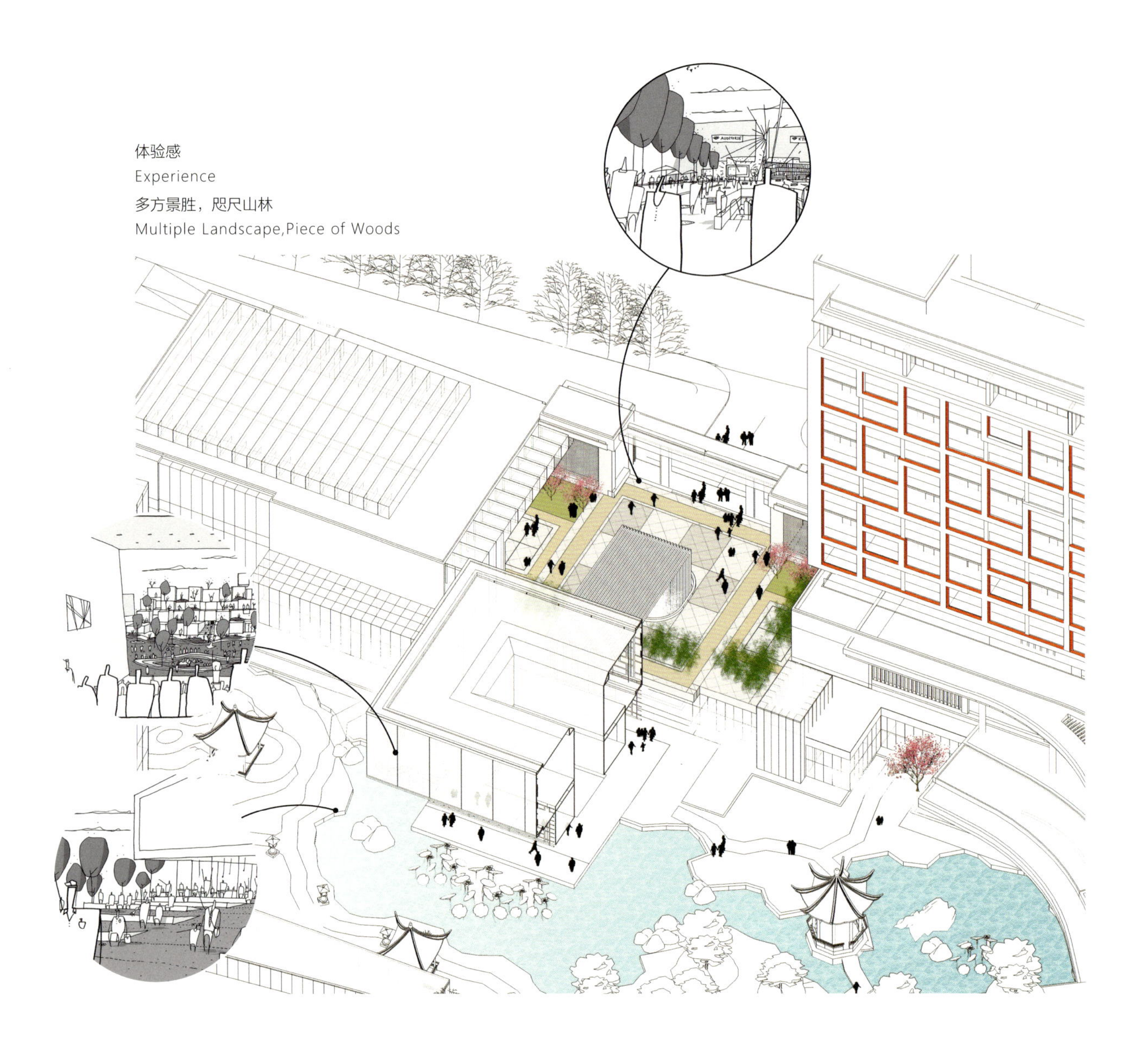
体验感
Experience
多方景胜，咫尺山林
Multiple Landscape,Piece of Woods

园林 · 路径
文化 · 体验
Garden· Path
Cultural ·Experience

园林文化的精髓在于路径的体验，在路径体验中感受最原味的园林文化

The Essence of Suzhou Garden is the Path, Experiencing the Most Original Garden Culture.

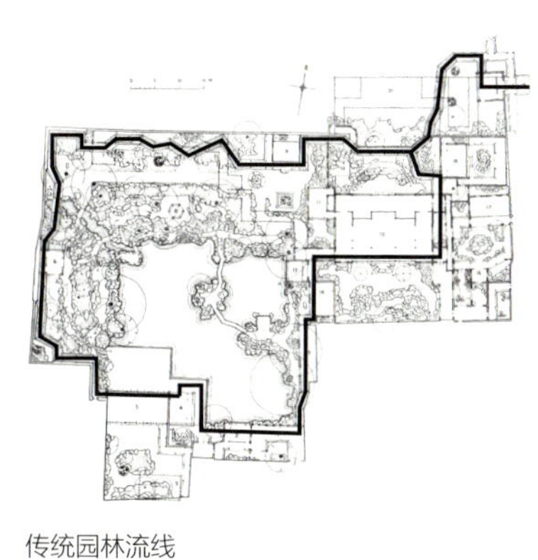

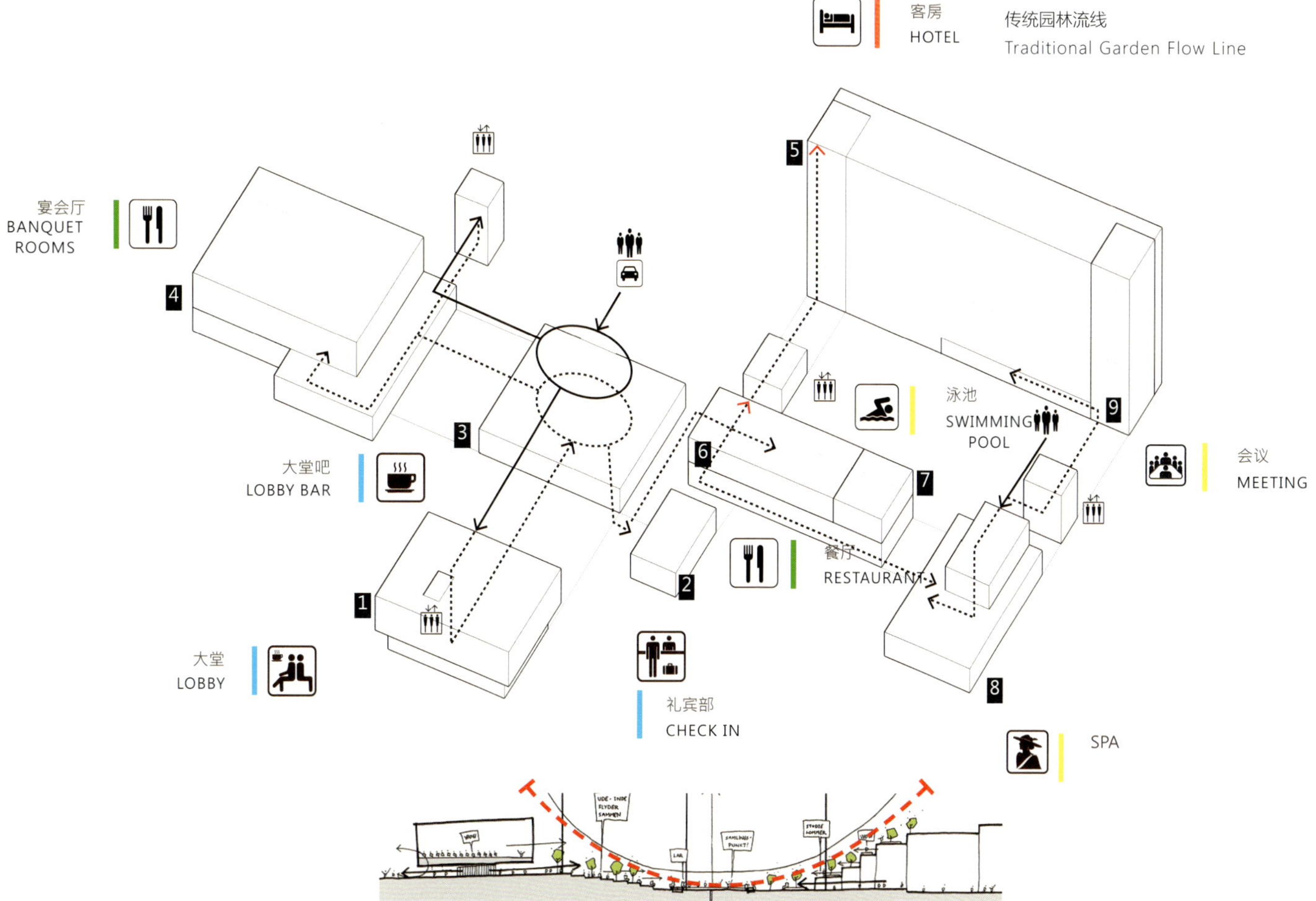

最终阶段模型组装过程 .2015 年 06 月
Final Stage Model Assembly Process.2015,06

如何凸显与现代奢华的关系？

奢华不仅在于外表的华美，更多的是体现在内涵的隽秀。而苏州园林的神韵正是这份设计中最秀美的内容。

苏州自古以园林的营造闻名遐迩，业主办公的总部也是一处精致的小园林。而造园的精髓正与现代酒店景观和空间设计的原理相同。嘉纳俗弃，精宜得体，将空间的故事在路径的转折跌宕中趣味悠长的展开，将构筑元素与文化结合以更文艺的方式展示，将人的行为模式与建筑形态结合并以画面叙事的方式描绘，这些正是酒店空间婉转而明朗的节奏韵味。于是在设计中，我们参考各大经典园林的作法和《园冶》的记载，对酒店大堂，内部中庭乃至延续的景观轴线都做了具体设计，将空间尺度、明度、维度和粘合度戏剧化的呈现出来，用现代的材料演示了古典空间的张力。

How to Highlight the Relationship with Modern Luxury?

Luxury is not only the appearance of the gorgeous, but also more is reflected in the connotation of the show. The charm of Suzhou garden is the most beautiful content of this design.

Suzhou is famous for its gardens. The headquarter of the proprietor's office is also an exquisite small garden. The essence of gardening is the same as that of the landscape of modern hotel and the concept of space design. Garnay abandonment, refinement, decentness, the story of space in the twists and turns of the path in a fun and unfolding, the combination of elements and culture will be displayed in a more literary manner. Human behavior patterns are combined with the architectural forms, depicted by the way of screen narration. These are the rhythms of the hotel space that is turning bright and clear. Therefore, in the design, we refer to the practices of major classical gardens and the *Records of Gardens and Smelts*. We have specifically designed the hotel lobby, the inner atrium, and even the continuation of the landscape axis to present the spatial scale, brightness, dimensions, and adhesion dramatically, using modern materials to demonstrate the tension of the classical space.

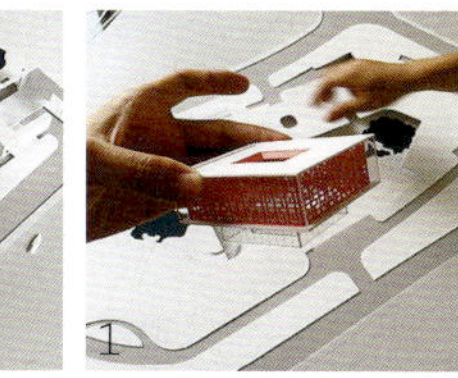

最终阶段模型组装过程 .2015 年 06 月
1.Final Stage Model Assembly Process.2015,06

交标前六小时 .2015.06.27.02：00 与模型独处的一点时间
2.Six Hours Before Submitting, Stay with the Model Alone 2015.06.27.02:00

同时，博古架式的立面设计也正是想凸显苏州传统木艺的精湛，在一虚一实的双重屏风式设计中，彰显文化魅力的镌刻。

从空间到建筑细节，苏州园林的渗透让这个现代的酒店有了浓郁的地方特色，有了历史沉淀的份量。正是有了这层内涵，我们的建筑才有了充满东方特色的独特艺术魅力。

At the same time, the design of the shelf-style facade is precisely to highlight the exquisite art of traditional wooden in Suzhou. In a virtual double-screen design, it highlights the charm of cultural charm.

From the space to the architectural details, the infiltration of Suzhou gardens has given the modern hotel a strong local character and a history of precipitation. It is the connotation that gives our buildings a unique artistic charm that is full of oriental features.

有高差的周期 / 南宁世茂五象国际中心

A Period with A High Difference / Shimao Five-Elephant International Center in Nanning

南宁

占地面积　: 24 398.46 平方米

建筑面积　: 233 348.98 平方米

建筑高度　: 160 米

设计时间　: 2017

Nanning

Floor Space : 24,398.46 m^2

Gross Floor Area : 233,348.98 m^2

Building Height : 160 m

The Time of Design : 2015

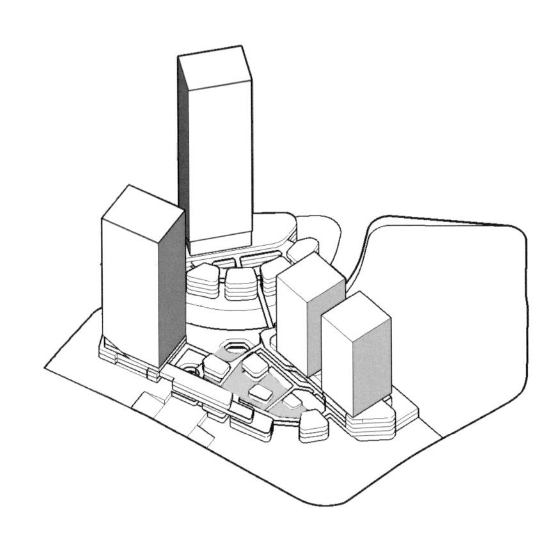

草坡上的花园商业

The Garden Business on the Slope

地产建筑师的专业素养不仅体现在建筑作品的成功，还有贯穿策划、设计、建造、营销每个环节的掌控力，清晰的逻辑和严谨的态度需要贯穿全程。南宁五象新区是一个功能复合的城市门户，自然与产业资源丰富。在政策推动下，周边布有多家知名开发商项目，恒大、绿地、万科等品牌楼板竞相扎根，其中毗邻的绿地中央广场都已建成。由于地块位于平乐大道发展轴的端点，作为新区的“南门户”，政府各部门关注度很高，而作为水石与世茂华南的首次合作项目，设计需要以高差异水准回应各方诉求。

项目在我们接手设计前，陆道设计团队已经进展了不短的时间，并得到政府首肯，甚至定为城市控规标准。在此标准中，塔楼的布局位置、最高塔楼的高度都已确定，而这与业主的营销判断与成本控制有着明显的差异。故而项目进展早期设计的复杂度最初主要集中在数据推演方面：土地出让条件中出售持有比、商业办公比的数据计算需要与分期数据、地上地下数据、计容数据动态计算考虑。因为地产设计的核心是市场、营销与品牌，也就意味着价值、去化和形象成为方案的核心，其中容量的数据模型和产品模型是基础。变量数据下规划方案均需平衡成本、营销到到最合理有利状态，数据统筹在地产类项目开发初期是设计必经的环节。

参数化思维对应变量

规划指标方面，出于鼓励城市立体空间向上发展，南宁当地规范规定建筑超100米部分只计一半容积率。同时由于场地有10米高差，在结合场地设计时，地上地下半地下部分也由于界定模糊，在多轮规划征询中由于规则不同得到的结果也不同。在局限地块中45%密度和30%绿化率的满足也增加了设计难度。同时，前后多部门甲方定位和产品诉求不统一，也造成产品价值的数据模型构建更加困难。

面对复杂多变的设计条件，我们从容量与产品的关系入手。针对超高层的可变容积率，测算不同高度和层高下产品配比和价值，以一期快速去化、二、三期高盈利率为基准选择最优化方案。同时，通过标准化设计控制产品级配。随着项目推进，针对业主方不同意见，我们就项目各期指标平衡、停车指标增倍后停车楼的建设等各方面都进行了专项研讨。

参数化思维的设计方式不仅为对应多元变量提供了解决途径，也为我们应对多轮反复式调整明确了设计方向。

The professional qualities of real estate architects are not only reflected in the success of architectural works, but also in the control of every aspect of planning, design, construction and marketing. Clear logic and rigorous attitudes are needed throughout the entire process. The New District of Nanning Wuxiang is a functional and complex city gateway with abundant natural and industrial resources. Under promotion of the policy, there are several well-known developer projects in the surrounding area. Brands such as Evergrande, Greenland and Vanke are racing to take root. The adjacent Greenland Central Plaza has already been built. Because the plot is located at the end of the development axis of Pingle Avenue, as the “south gateway” of the new district, government departments pay high attention to it. As the first cooperation project between Water Stone and the South China of World Trade, the design needs to respond to the demands of parties with a high level of difference.

Before design of the project was took over by us, the land route design team had made continuous progress and received government approval and even targeted standards of urban control. In the standard, the layout of the tower and the height of the tallest tower has been determined, which is obviously different from owners' marketing judgment and cost control. So the complexity of the early design of the project was initially focused on data deduction: the calculation of sale-to-hold ratio in the land transfer condition, the calculation of commercial office ratio, and the dynamic calculation of staging data, ground-floor data, and volume-volume data. Because the core of design of real estate are markets, marketing and branding, which means that value, desaturation, and image are the core of the program. The model of data and product of capacity are the foundation. Under variable data, the planning scheme needs to balance the cost and marketing to the most reasonable and favorable state. Data co-ordination is a necessary part of the design in the initial stage of the development of projects of real estate.

The Parametric Thinking Matches Variables

In terms of planning indicators, it is encouraging the urban three-dimensional space to develop upwards. The local regulations in Nanning stipulate that the building exceeds 100m in volume and only accounts for half of the volume rate. Meanwhile, due to height difference of 10m in the site, the semi-underground part of the ground and underground is also blurred because of the difference in the design of the site. The result obtained in multiple rounds of planning consultation is different due to different rules. The satisfaction of 45% density and 30% greening rate in the confined land also increases the difficulty of design. Meanwhile, the orientation of multi-departmental party A and appeals of products are not uniform before and after, which also makes it more difficult to construct a data model of product value.

Facing complex and ever-changing conditions of design, we begin with the relationship between capacity and products. For the variable volume ratio of super high floors, the ratio and value of products at different heights and floor heights are measured, and the optimization plan is selected based on the rapid turnaround of Phase I and the high profitability of Phase II and Phase III. Meanwhile, product grading is controlled through standardized design. As the progress of project, we have conducted special discussions on various aspects of owners’ opinions in terms of balance of various indicators of the project, and the construction of parking buildings after the parking index has been doubled.

The approach of paradigm design not only provides a solution to multiple variables, but also clarifies design direction for us to deal with multiple rounds of repeated adjustments.

五象新区门户、高差地形，超高层建筑群、
复杂动态数据运算、睿智创新、
全天候花园式商业街区。

Wuxiang New District Portal, Terrain of Difference of Height, high-rise buildings,
Complex and Dynamic Data Operations, Intelligent Innovation,
All-weather and Garden-style Commercial District

22°45′
27.13″ N
108°22′
35.57″ E

界面分析
Interface Analysis

主次界面
Primary and Secondary Interface
依据人气来测定时间轴
Measuring Timeline Based on Popularity
干道展示面 445 米
Main Road Display Surface 445m
城市可观察视角约 180°
The Observable Angle of the City is About 180°

主要界面
Main Interface
次要界面
Secondary Interface

模拟视线体验设定点
Analog Line of Sight Experience Set Point
4：00-10：00

A 级界面时长	B 级界面时长
A Class of Interface	B Class of Interface
约 13.5 秒	约 4.5 秒

模拟视线体验分段
Analog Line of Sight Segmentation
18/6=3s

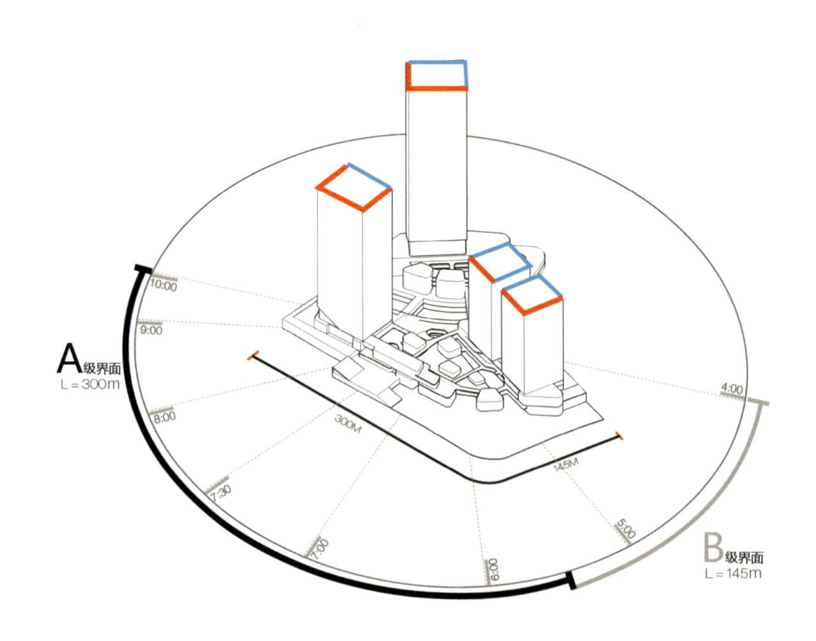

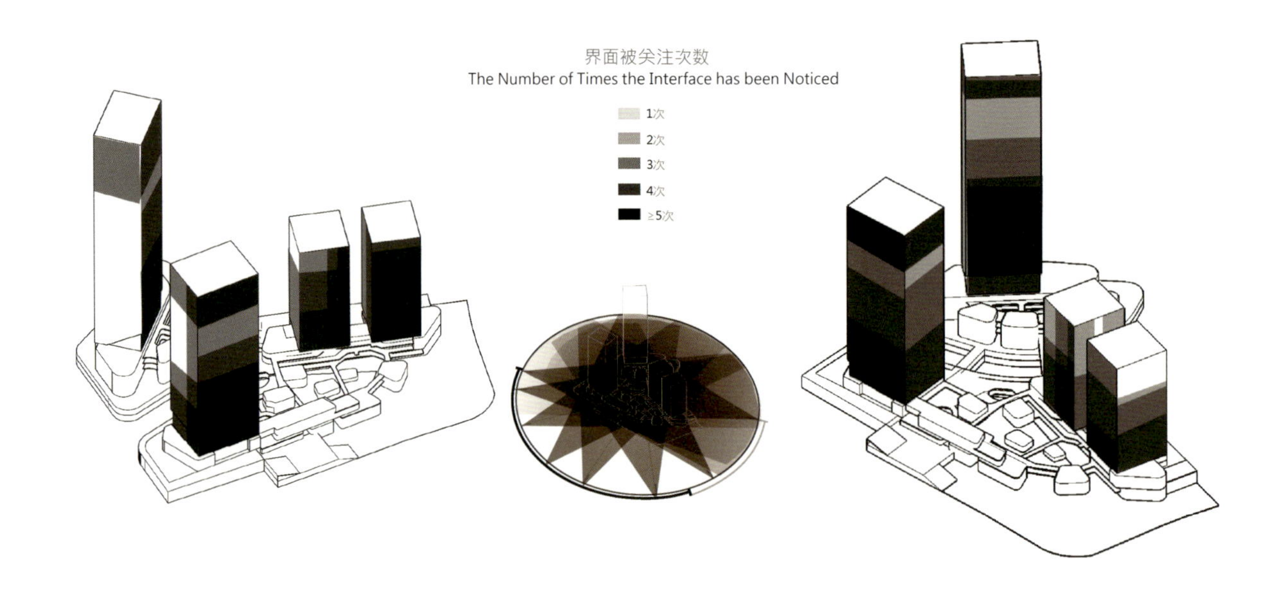

在花园绿坡上散步的商业

除了塔楼布局受约束、数据模型建立复杂外，10 米高差的场地约束是这个项目的最大难点之一，尤其要在门户地段超高层建筑裙房部分营造一处品质较高的商业街区。3 万平方米的容量，如果按常规沿街商业模式布局，难以塑造强烈的话题感。为了匹配门户位置的地段价值，我们将其设计为一处城市客厅，以原地形的巨大高差形成空间体验的张力，以生态自然回应当地绿色环境，打造一处休闲散步的花园绿坡。

结合 10 米高差和微地形设计，场地形成三个标高，达到三首层效果，极大提升营销价值。在商街首层，半围合的广场中绿色草坡上散布独栋商业，草坡下则平接车库。草坡本身既可作为运动、艺术等多主题的展示场所，也为城市提供了多生态休闲空间，成为一大话题。

Business that Walks on Green Slope of the Garden

In addition to the constrained layout of the tower and the complex establishment of the data model, the site constraint of the elevation difference of 10m is one of the greatest difficulties in this project, especially the creation of a high-quality commercial block in the portal section of the super high-rise building. With a capacity of 30,000 square meters, it is difficult to create a strong sense of topic if it is arranged according to a conventional business model along the street. In order to match the value of the location of the gateway, we designed it as a city living room, forming the spatial experience of the tension with the height difference of the original terrain, responding to the local green environment with the ecology, and creating a leisurely walk garden green slope.

In combination with the 10m elevation difference and micro-topography design, the site forms three elevations and achieves three levels of effectiveness, which greatly enhances marketing value. On the first floor of Shang Street, a half-enclosed square takes a walk on a green turf slope and separates a commercial. Grassland itself can be used as a venue for multiple themes such as sports and art. It also provides a multi-ecological leisure space for the city, which has become a topic.

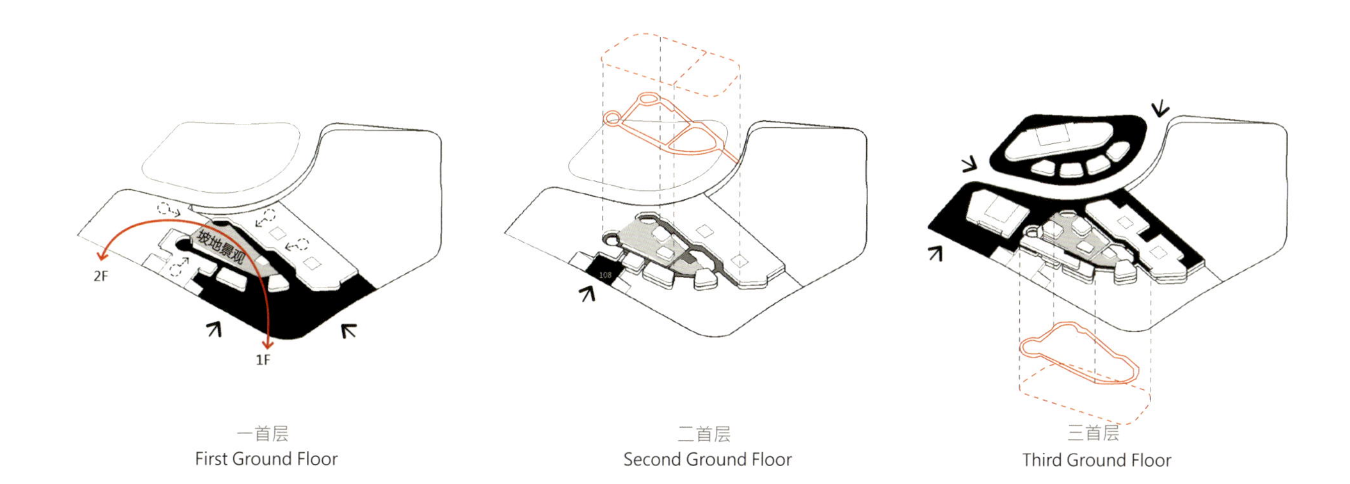

一首层 First Ground Floor

二首层 Second Ground Floor

三首层 Third Ground Floor

商业“点状分布”总平面

The Ichnography of "Spot Distribution" of Commerce

草坡上的“花园街区”

Garden Block on Grassy Slope

控制窗墙比及标准段建造

Control Window Wall Ratio and Standard Section Construction

100m 公寓标准段分析

Analysis of Standard Section of 100m Apartment

标准段组合形式：

重复

Standard Section Combination Form:

Repeat

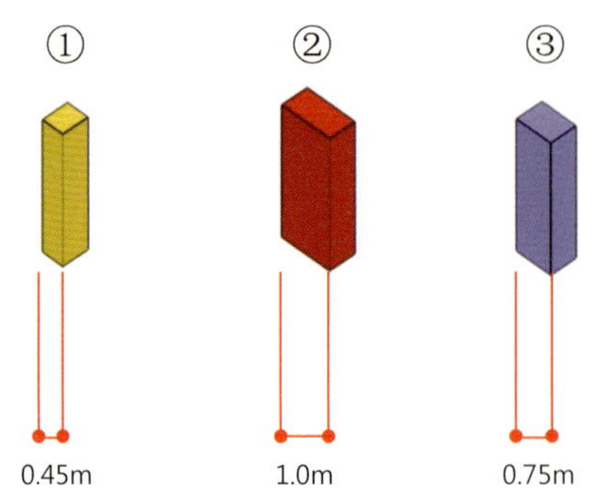

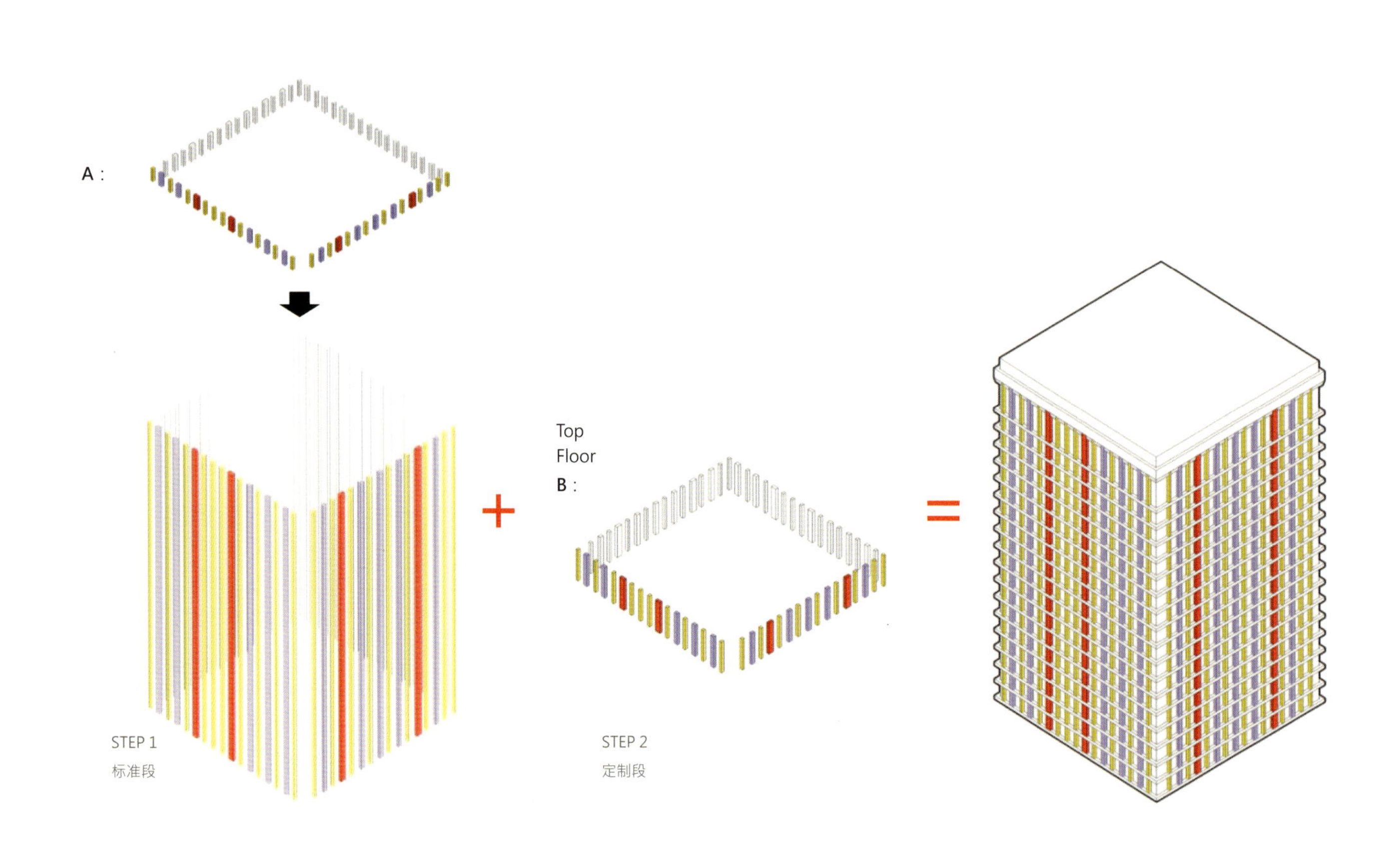

160m+200m 塔楼标准段分析

Analysis of Standard Section of 160m+200m Tower

标准段组合形式：

并列

Standard Section Combination Form:

Parallel

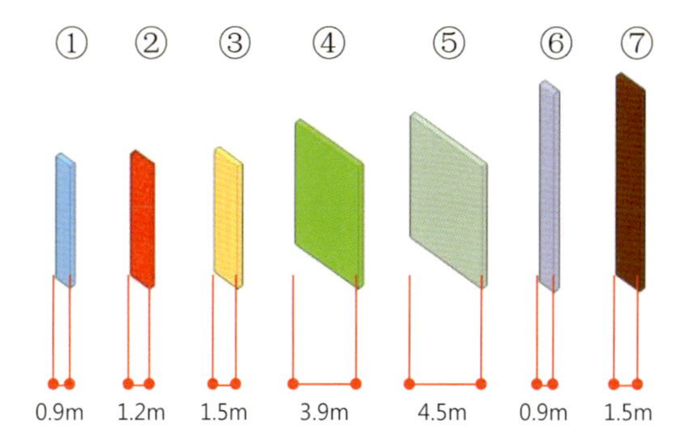

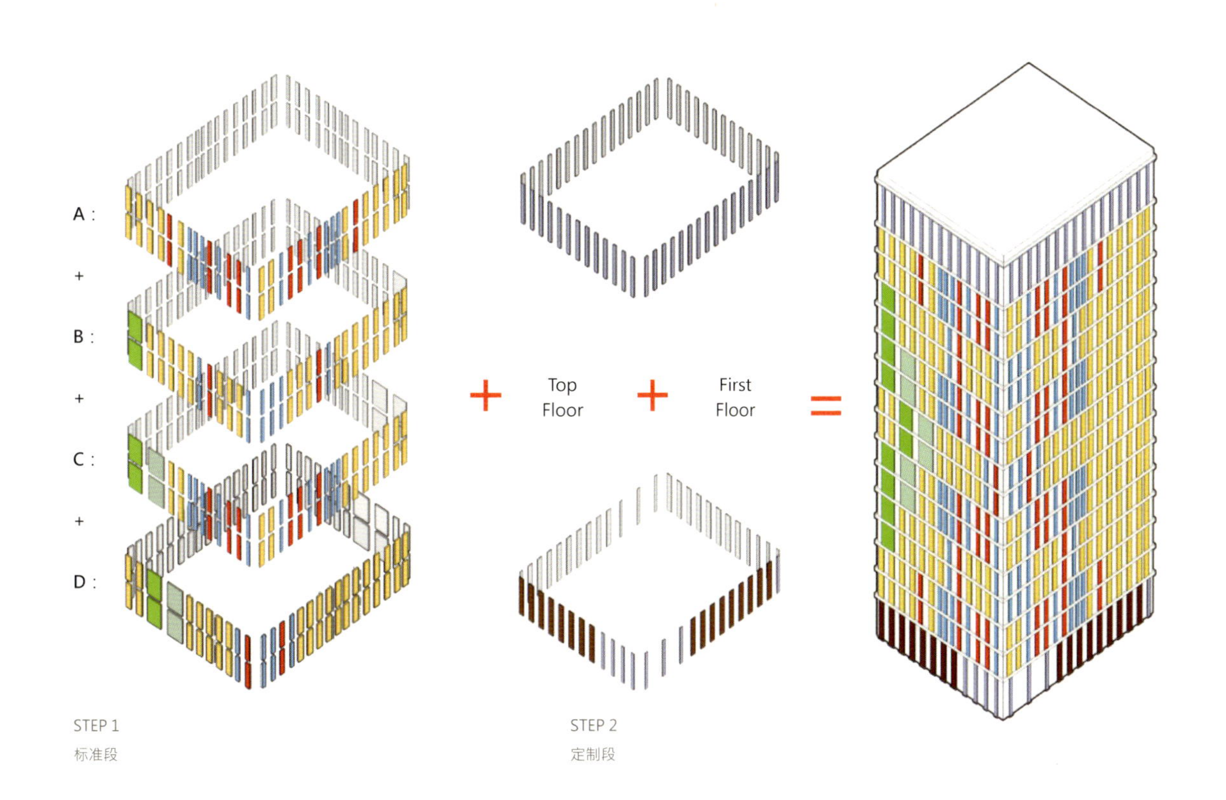

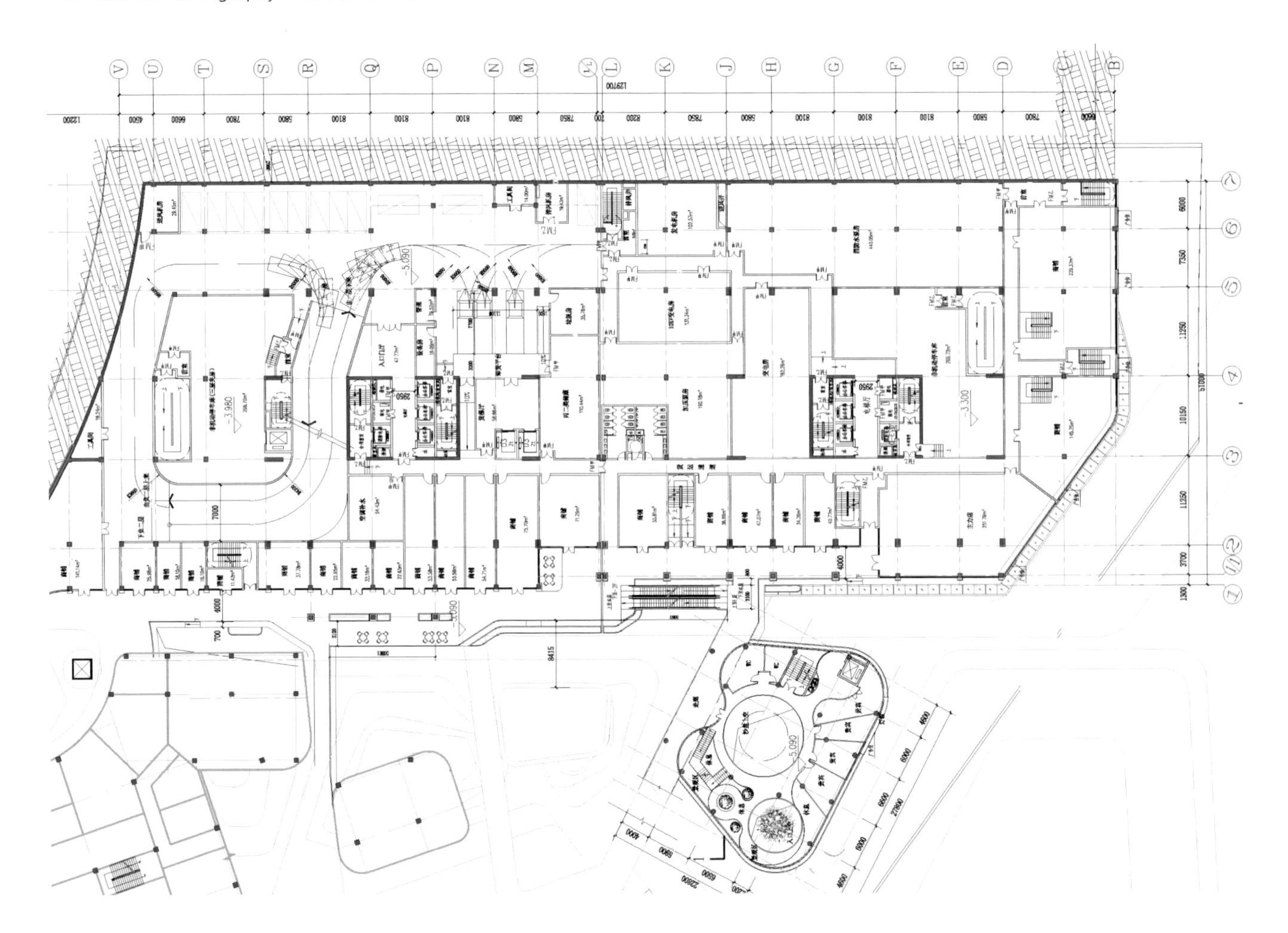

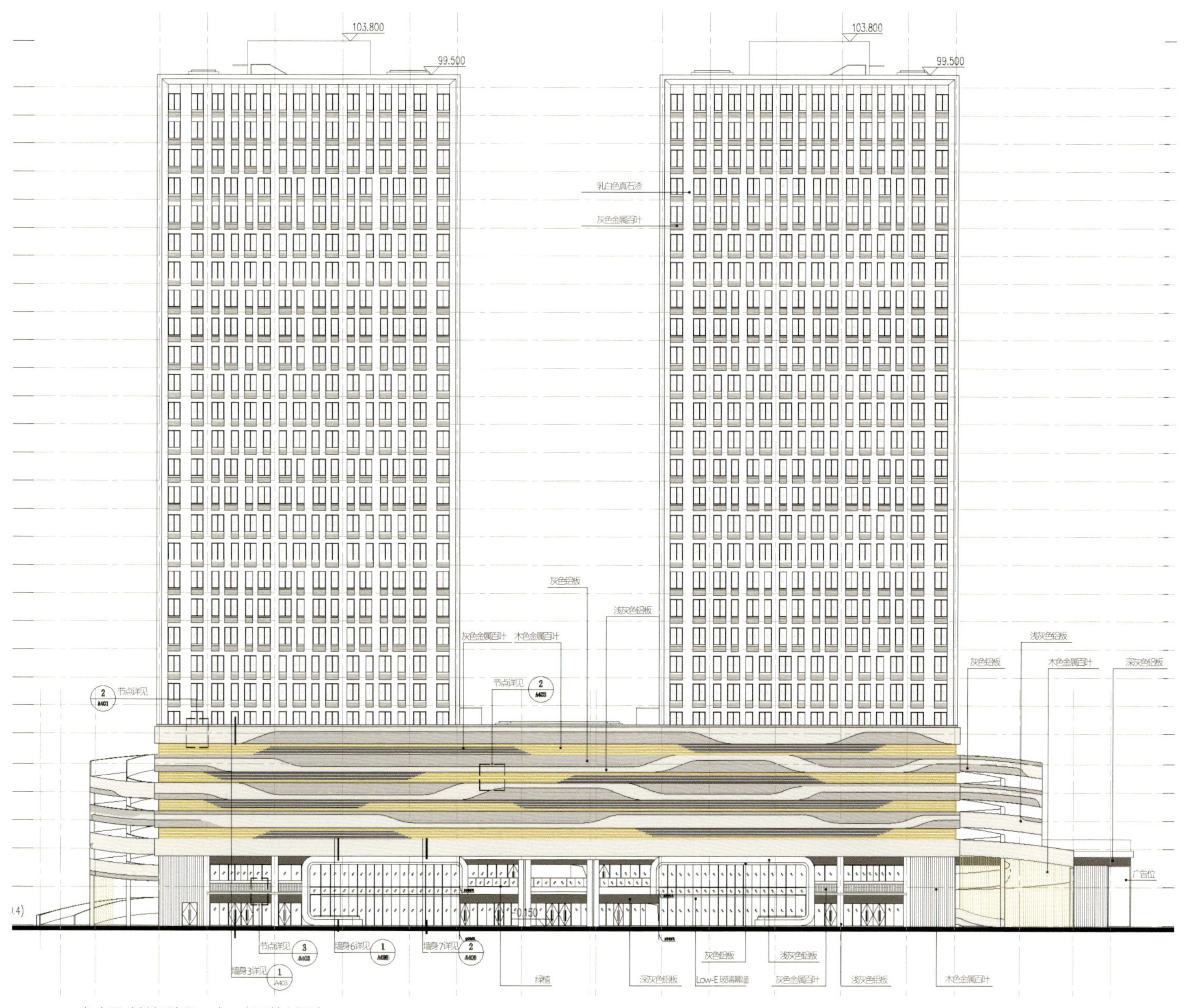

东立面（扩初阶段，成一立面控制图）

East Facade (Elevation Control Chart by One Studio)

有声的默剧 / 上海崇明长岛游艇码头

The Phonic Dumb Show / Chongming Long Island Marina in Shanghai

上海

占地面积 ： 78 185 平方米

建筑面积 ： 76 255 平方米

容积率 ： 0.98

设计时间 ： 2014

Shanghai

Floor Space : 78,185 m^2

Gross Floor Area : 76,255 m^2

Plot Ratio : 0.98

The Time of Design : 2014

绿地崇明新长岛
Greenland Chongming Long Island

芭蕾舞舞者只要还在坚持旋转，心中的舞台就不会散场，即使灯光比默剧还要沉寂。这是绿地 2014 年充满战略野心的一个项目。当时，绿地正想赶超万科，成为上海乃至全国第一大开发商，然而缺乏极具影响力的特色项目。于是绿地结合时势，积极开拓国际化视野，试图在上海创造出有世界级影响力的旅游度假项目，以此吹响向市场龙头进军的号角。绿地长岛就在这样的背景下应运而生，而我们参与竞标的这个项目也就在长岛的中央。

长岛是一处紧邻崇明岛的人造岛，扼长三角北入海口，南依崇明岛，北面江苏启东、海门，总占地约 15.4 平方公里，规划面积约 2.35 万亩，全长达 17.5 千米。绿地请荷兰设计公司将其规划为四段五节点，希望打造成有国际影响力的集养老、度假、旅游为一体的休闲胜地。我们设计的核心码头区在其中央区域，环境资源优秀，业主也极为重视，组织了 Aedas、UA、栖城、山水秀和我们水石共五家公司在三周内投标设计。

我们当时作为一支缺乏旅游地产设计经验的团队全力以赴，调研了大量国际一线旅游开发项目，探讨了多轮规划和建筑方案，直至最后一轮都得到了业主的充分认可。虽由于各方原因，未能中标实现，但仍给对方留下了深刻的印象，为水石与绿地在包括长岛的多处深化合作创造了新的契机。

As long as the ballet dancers are still spinning, the stage in the heart will not be dispersed, even if the lights are silent than the mimes.
This is a project of Greenland that is full of strategic ambition in 2014. At that time, Greenland was trying to catch up with Vanke, becoming the largest developer in Shanghai and even in the country. However, it lacked influential features. Greenland actively explored an international perspective and attempted to create a world-class project of tourism and holiday in Shanghai in combination with the current situation, which blew the clarion call to enter the market. Greenland Long Island came into being in this context. The project we participated in bidding for is also in the center of Long Island.

Changdao is an artificial island next to Chongming Island. It is located in the north of the Yangtze River Delta, Chongming Island in the south, and Qidong and Haimen in the north. It covers an area of approximately 15.4 square kilometers and covers a planned area of approximately 23,500 acres. The total length is 17.5 kilometers. Greenland asked Dutch design company to plan it as a four-node and five-node node. It hopes to create a leisure resort that integrates pensions, vacations and tourism with international influence. The core dock area we designed was in its central area, with outstanding environmental resources. And proprietors attached great importance. The five companies of Aedas, UA, Qicheng, Shanshuixiu, and W&R were invited to bid for the design within three weeks.

At that time, we, as a team lacking experience in the design of tourism and real estate, went all out to investigate a large number of international first-line tourism development projects, discussed multiple rounds of planning and construction programs, and were fully approved by the owners until the final round. Although due to various reasons, the the bid was not achieved. But it still left a deep impression on the other party, creating an opportunity for the cooperation between W&R and Greenland in many fields including Long Island.

崇明品质岛居、生态湿地、城市近郊
商、住、游、购一站式度假体验
核心主题区的风情设计

Residence of High Quality of Chongming Island, Ecological Wetland, Urban Suburbs,
One-stop Holiday Experience of Business, Accommodation, Travelling and Purchasing ,
Thematic Design of the Area of Core Theme

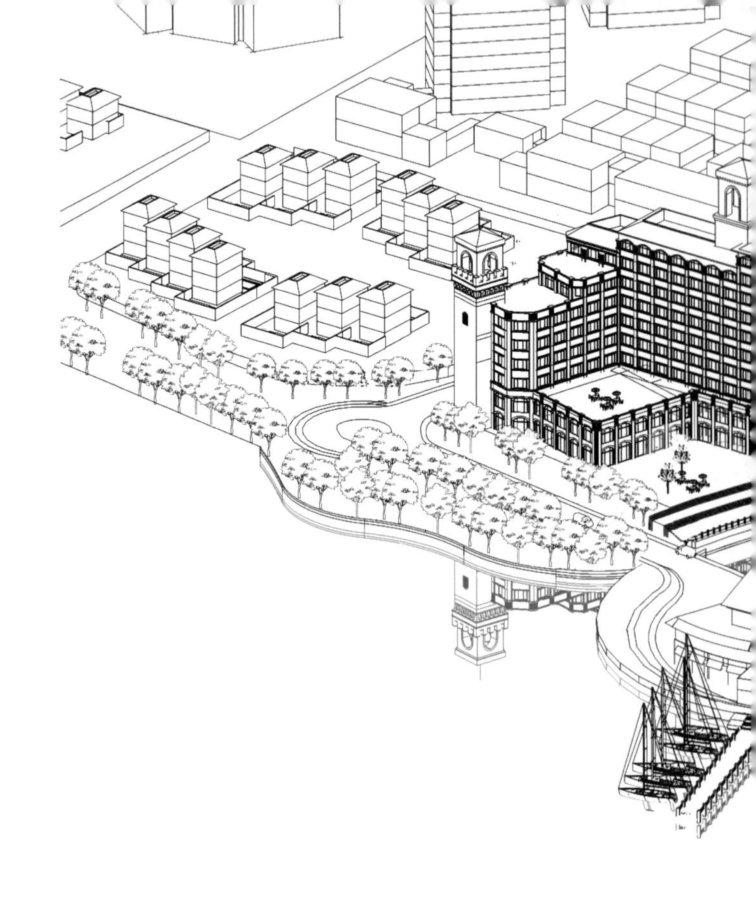

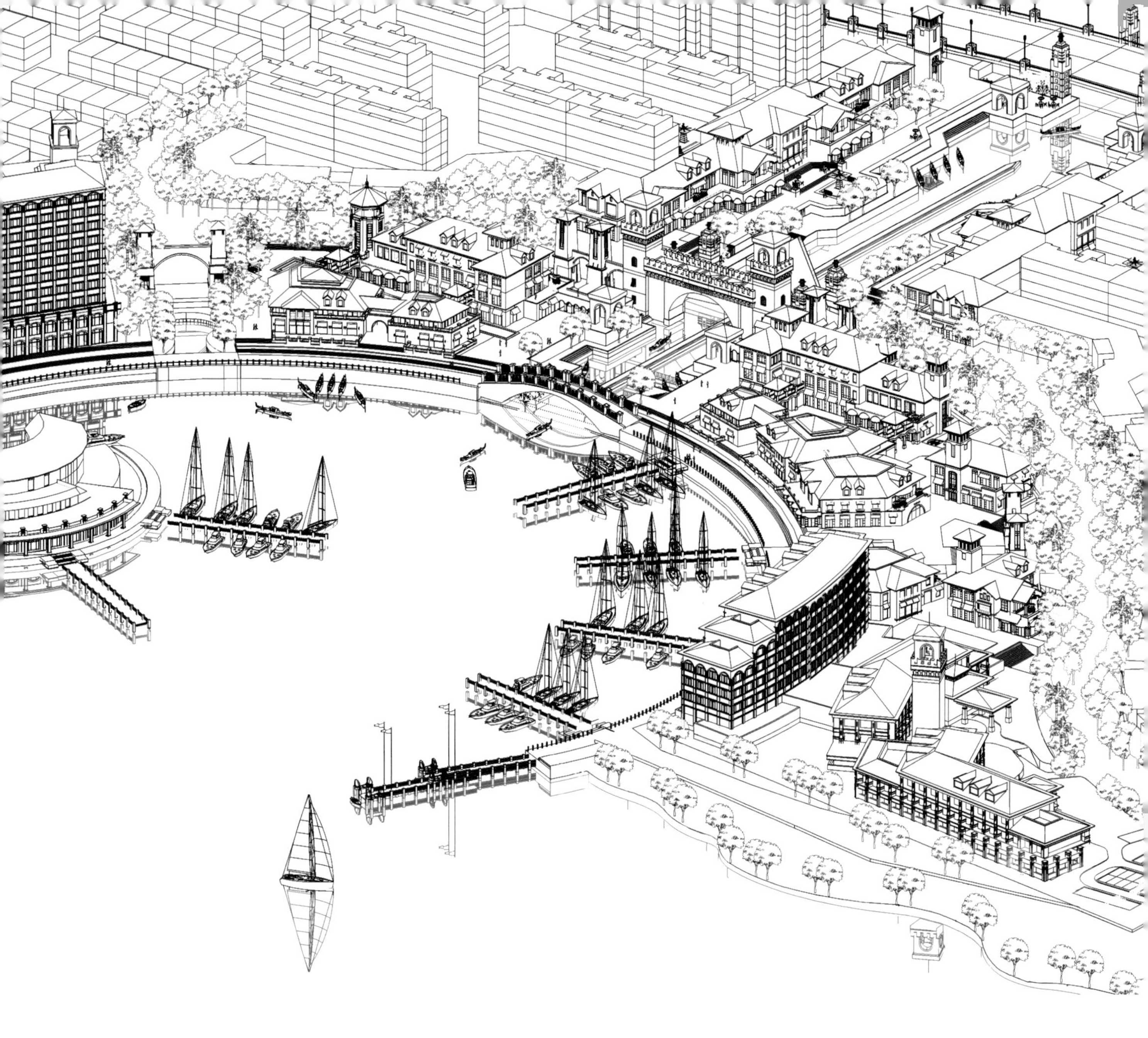

31°24′
25.00″ N
121°41′
33.12″ E

以欧洲港湾小镇肌理展现度假风情

Show the Holiday Style with the Texture of European Harbor Town

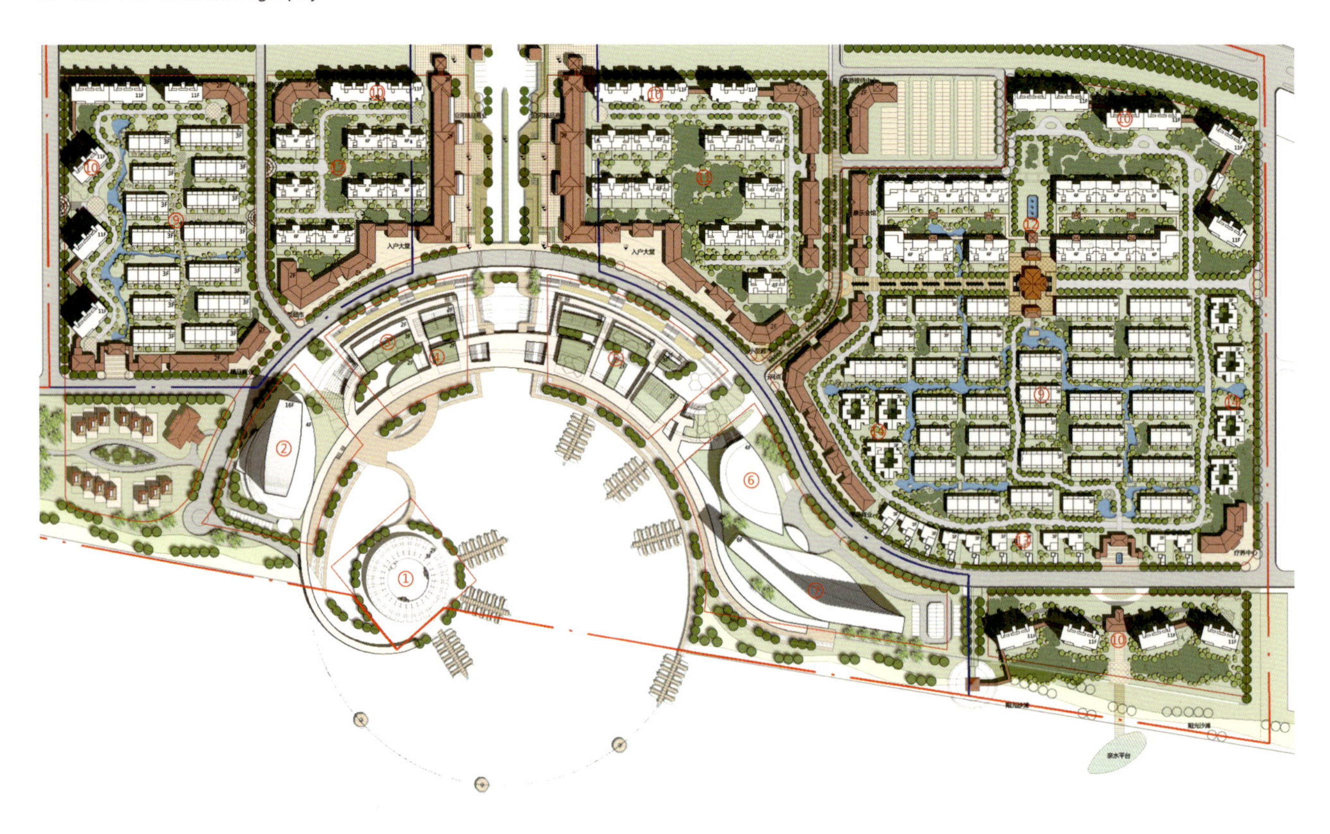

一湾风情的海岸

作为有国际影响力的度假休闲胜地，营造海边的风情是项目的核心。在这片港湾，我们需要布置一座五星级酒店、一座四星级酒店，一栋疗养中心，一段风情商业街和一处特色的游艇码头俱乐部。设计参考了多处世界一流度假胜地案例，在其中考量业态、场地、建筑与水的关系，描摹海边度假的生活状态。我们还参考了大量欧洲小镇中原汁原味的街区风情，揣摩真实空间中的尺度、材质、肌理、符号与文化与人的黏性，穿透建筑表皮和景观细节，感受当地环境中特殊的声音、气味和热度。

在多案例参考、多维度考量后，我们提出了两稿设计方案。一稿类似于海港，设计以欧洲港湾小镇肌理展现度假风情。另一稿则更偏现代，并以多退层的形式体现亲水的特质。两稿设计都整体呈现了象牙湾式的流线形态，码头鲜明的完形设计也给业主留下了极深的印象。为了适应绿地节奏，我们将体块比例的细致推敲结合拼贴式的快速呈现，从而实现了短周期内高效率的设计操作。

An Amorous Coast

As a holiday resort with international influence, creating a seaside style is the core of the project. In this bay, we need to arrange a five-star hotel, a four-star hotel, a rehabilitation center, a style commercial street and a special yacht marina club. The design refers to a number of world-class resort cases, in which the relationship between business, venues, architecture and water is considered, describing the living conditions of seaside resorts. We also refer to the authentic street style of a large number of European towns and try to figure out the dimensions, textures, textures, symbols, cultures and people's viscous in real spaces, penetrating the architectural skin and landscape details to experience the special sound, odor and heat in the local environment.

After references multiple cases and multi-dimensional considerations, we proposed two draft designs. The manuscript is similar to the harbor. It is designed to show the holiday style with the style of the European Harbour town. The other draft is more modern and reflects the hydrophilic nature in a multi-layered form. The design of the two manuscripts presented the streamline shape of the Ivory Cove as a whole. The distinctive design of the terminal also left a deep impression on the owners. In order to adapt to the rhythm of the green space, we combined the detailed scrutiny of body mass ratios with the rapid presentation of collages to achieve efficient operations of design in a short period of time.

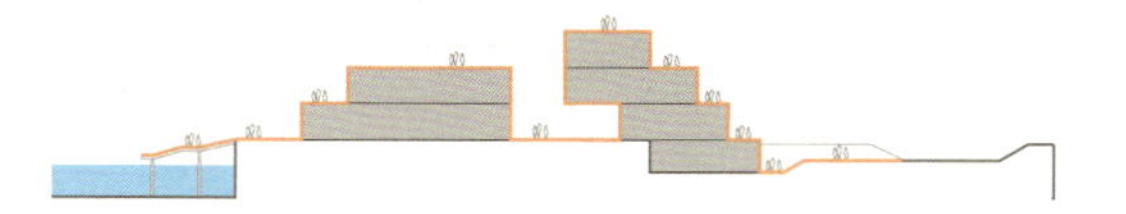

沿湖木栈道观景休闲、内街进人尺寸、商铺双首层

Lakeside Boardwalk Sightseeing Leisure

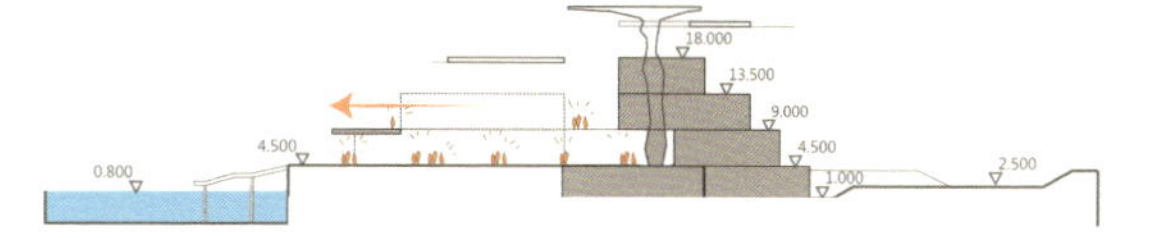

开放的广场给商业活动提供了场地，是商街人流的汇合点

The Site is the Confluence of People in the Business Street

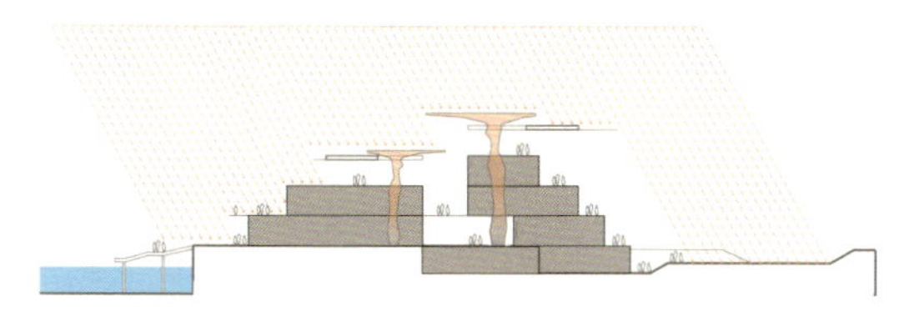

高品质屋顶灰空间提高商铺吸引力，增加商业价值

Roof Grey Space Increases Business Attraction and Ancreases Commercial Value

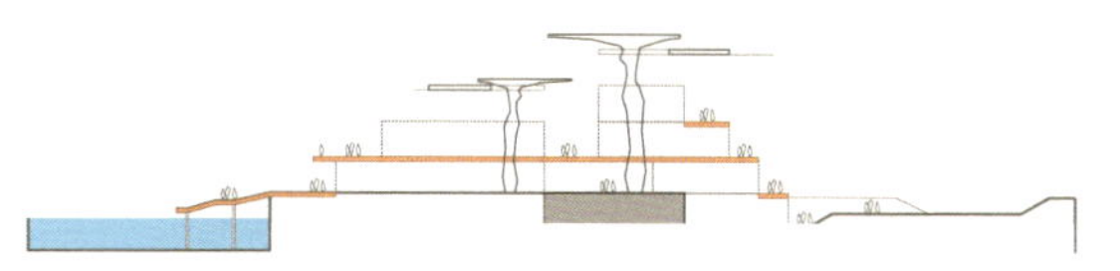

贯通商街的多层复合流线带来宽景的视觉享受，增加商业价值

Complex Streamlines Bring Visual Enjoyment and Increase Commercial Value

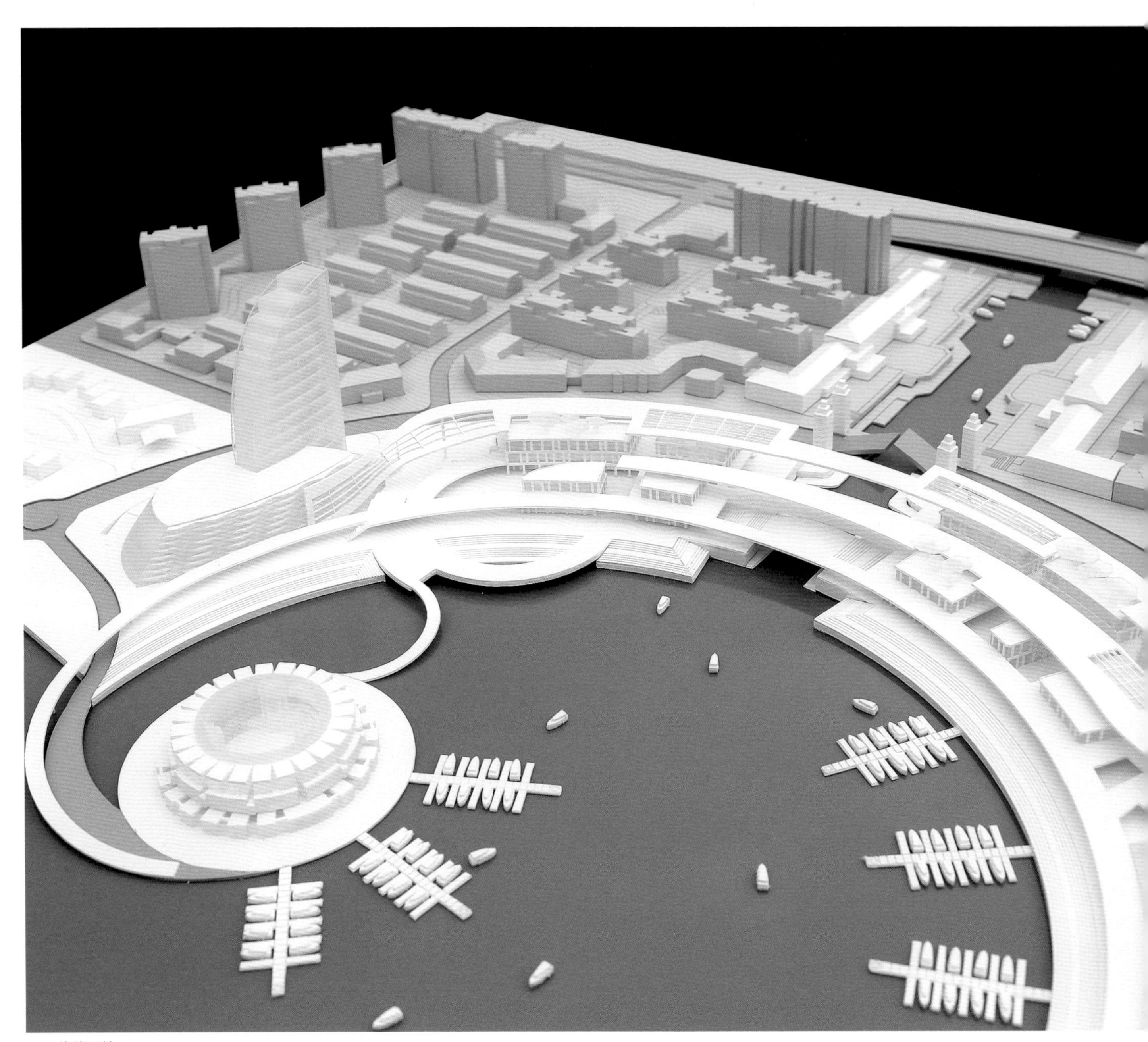

海岸风情

Local Conditions and Customs of the Coast

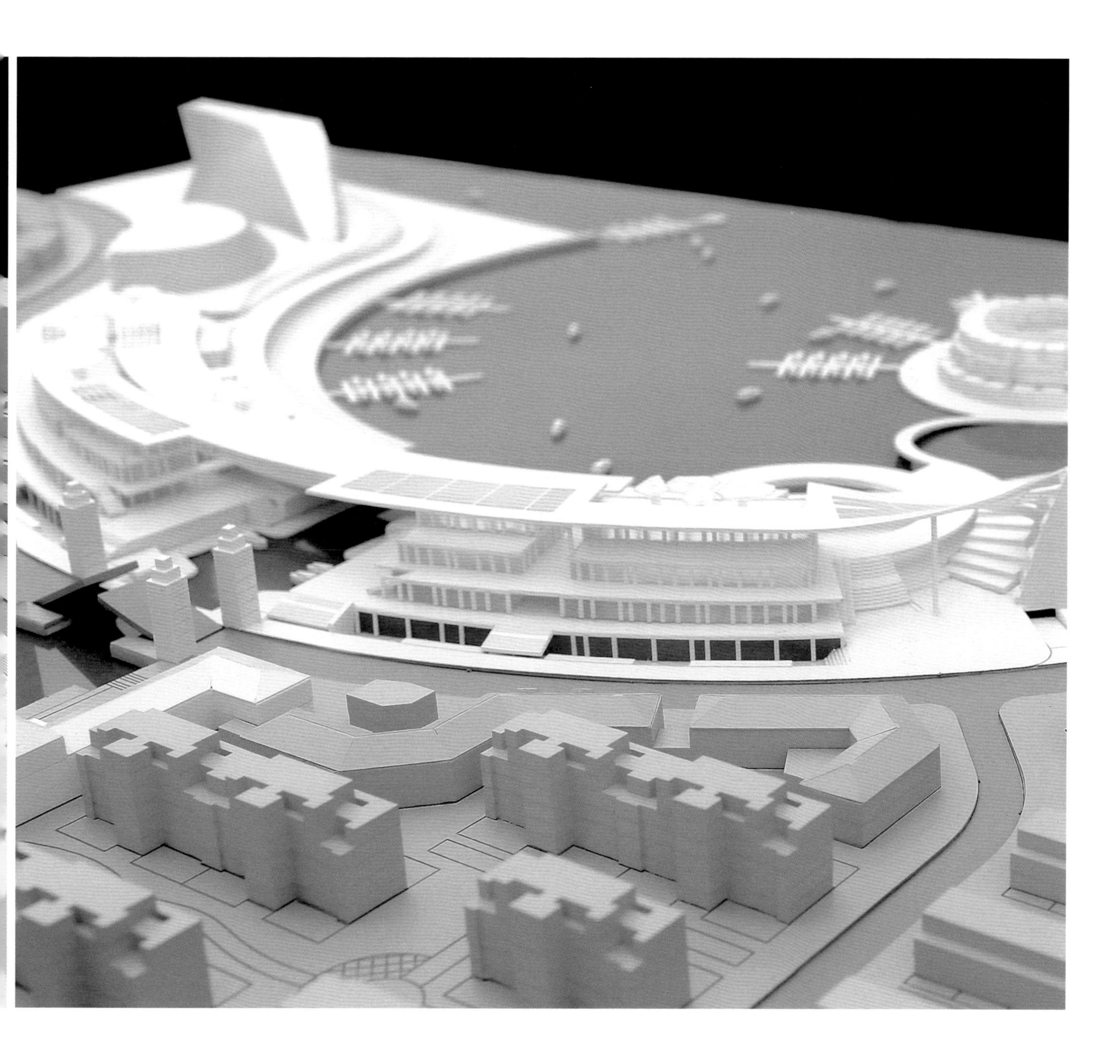

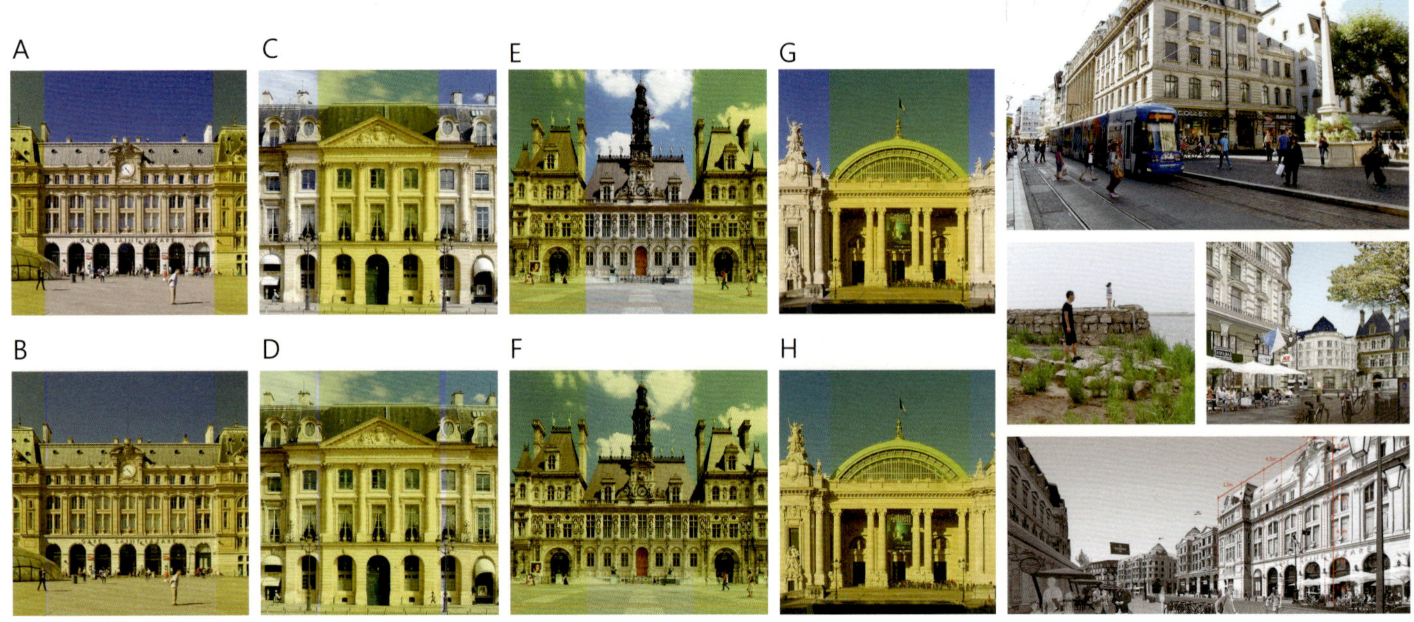

融合真实比例的标准段

Standard Segments that Fuse True Proportions

唱罢才发现是场默剧

本以为将是一轮结束的投标方案，而最终我们持续参加了四轮，每轮过后留给设计的时间也越来越短。最后，我们也仅剩一名对手。在最后一轮，业主甚至希望在 2 天内，我们能提出一稿正统法式建筑的设计方案。在如此极短的时间内，呈现比例精准、细节完美的成果是几乎不可能的。于是我们研究了全新的表达方式，以六成的时间找素材，四成的时间建体块和贴图，快速呈现了最终效果。这样的效率让业主也很惊讶，可是经历最终犹豫后，还是把设计权交给了另一家。事后才知道，这是竞标开始前就定下的结局，而我们只是配合了四轮深化。

这就是现实。面对这样的现实，我们需要的不仅仅是设计素养，还需要成熟的心理素质。心态是呈现给人的最鲜明的姿态，而耐心也是我们为这场默剧献上的最诚恳的祝福。

After Finishing Singing, It Was Discovered that It Was a Mime.

We thought that it would be a round of proposal of bidding. But in the end we continued to participate in four rounds. And the time left for designing after each round was getting shorter and shorter. Finally, we only have one opponent left. In the last round, the proprietors even hoped that we could propose a scheme of designing for orthodox French architecture within 2 days. In such a short period of time, it is almost impossible to present the achievement that has precise proportion and perfect details. So we studied a new way of expression, looking for materials with 60% of the time, building blocks and textures with 40% of the time, and quickly showing the final results. This efficiency has surprised the owners. But after hesitantly hesitating, he left the right of designing to another. We know afterwards that it was the outcome that was set before the beginning of the bidding. We only cooperated with the four rounds of deepening.

This is the truth. Facing this kind of reality, we need more than just the literacy of designing, but also mature psychological qualities. The state of mind is the most vivid gesture that is presented to people. Patience is our most sincere blessing for the silent drama.

法式风格 SU 快速表达

Rapid Expression of French Style SU

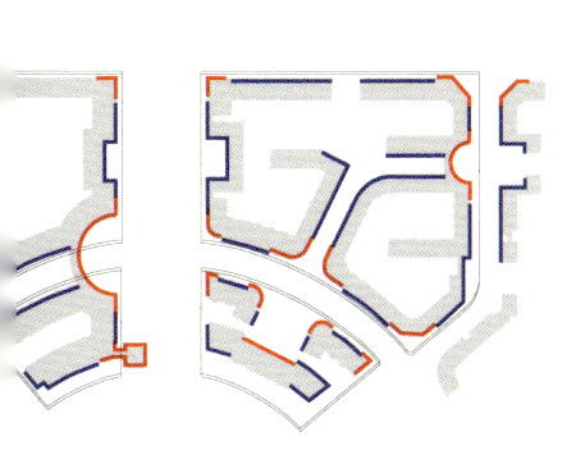

听得见，摸不着 / 兰州榆中沿川文园

Audible But Impalpable / Yuzhong Yanchuan Garden in Lanzhou

兰州

占地面积 ：45 873 平方米

建筑面积 ：58 000 平方米

容积率 ：1.26

设计时间 ：2016

延续文脉

这是正和集团的情怀力作。这是一个由兰州发迹的业主回馈家乡的故事。这是将高庙的福祉带回到坊间百姓的项目。这是一个在沿川湖边的最质朴文化憧憬。甚至项目尚未启动，业主已耗巨资整治山林，2 万余株树陆续将黄土变为绿野，40 余米湖底淤泥整治一空，所有的一切都是为了完一个回归山林的梦。

我们用笔尖不停摹画当地的山野与村庄肌理，翻阅众多有关民居的文献资料，多次去往当地考察调研，陪同业主与村民交流，在乡音与风物中体验这里生活的温度。无论是山峦起伏中的颠簸，还是饱经沧桑的市庙那边传来的铜响，都是当地最自然的生活。我们在维系文脉的基础上改造布局，让每家每户在干净而开敞的院子中得到安心的居住环境，让孩子们在儒家学堂般的场所中领略传统文化，让集市、祠庙、广场的游览路线与田野的风光融合，让偶然发现这块乡土的旅客品尝最兰州的滋味。十年树木，百年树人，为了让这份最本色的乡土韵味回馈生活而超越轴线组团的图纸规划，我们的投入远远超过这些。

Lanzhou

Floor Space : 45,873 m^2

Gross Floor Area : 58,000 m^2

Plot Ratio : 1.26

The Time of Design : 2016

Continuation of Cultural Venation

This is the spirit of the Zhenghe Group. This is a story of the owners of Lanzhou who gives back to their hometown. This is a project that brings the well-being of Gaomiao back to the people in the area. This is one of the most unadorned cultures along the shore of Yanchuan Lake. Even the project has not yet been started, the owners have spent huge sums to rehabilitate the forest. More than 20,000 trees have gradually turned the loess into a green field and more than 40 meters of mud on the lake floor have been rectified. All of this is to complete the dream of returning to the forest.

We used the tip of the pen to draw paintings of the local mountains and villages, read numerous literatures about residents, went to the local area for investigations several times and accompany proprietors to communicate with the villagers, experiencing the temperature of life in the township and the wind. Whether it is bumps in the mountains, or the copper tingle that has passed through the vicissitudes of the temple, it is the most natural life in the area. Based on the maintenance of the cultural venation, we have transformed the layout so that each household gets a safe living environment because of a clean and open courtyard, allowing children to comprehend the traditional culture in a Confucian school-like setting, letting the market and temples, the tour of the square integrate into the scenery of the fields to make the visitors who stumbled upon the local taste the most characteristic Lanzhou. Ten years of trees, a hundred years people. In order to give this most authentic native flavor back to life and go beyond the drawing plans of the axis group, our investment goes far beyond these.

文化性定制乡村改造
Rural Reconstruction of Cultural Customization
35°40′
27.97″ N
104°12′
21.84″ E

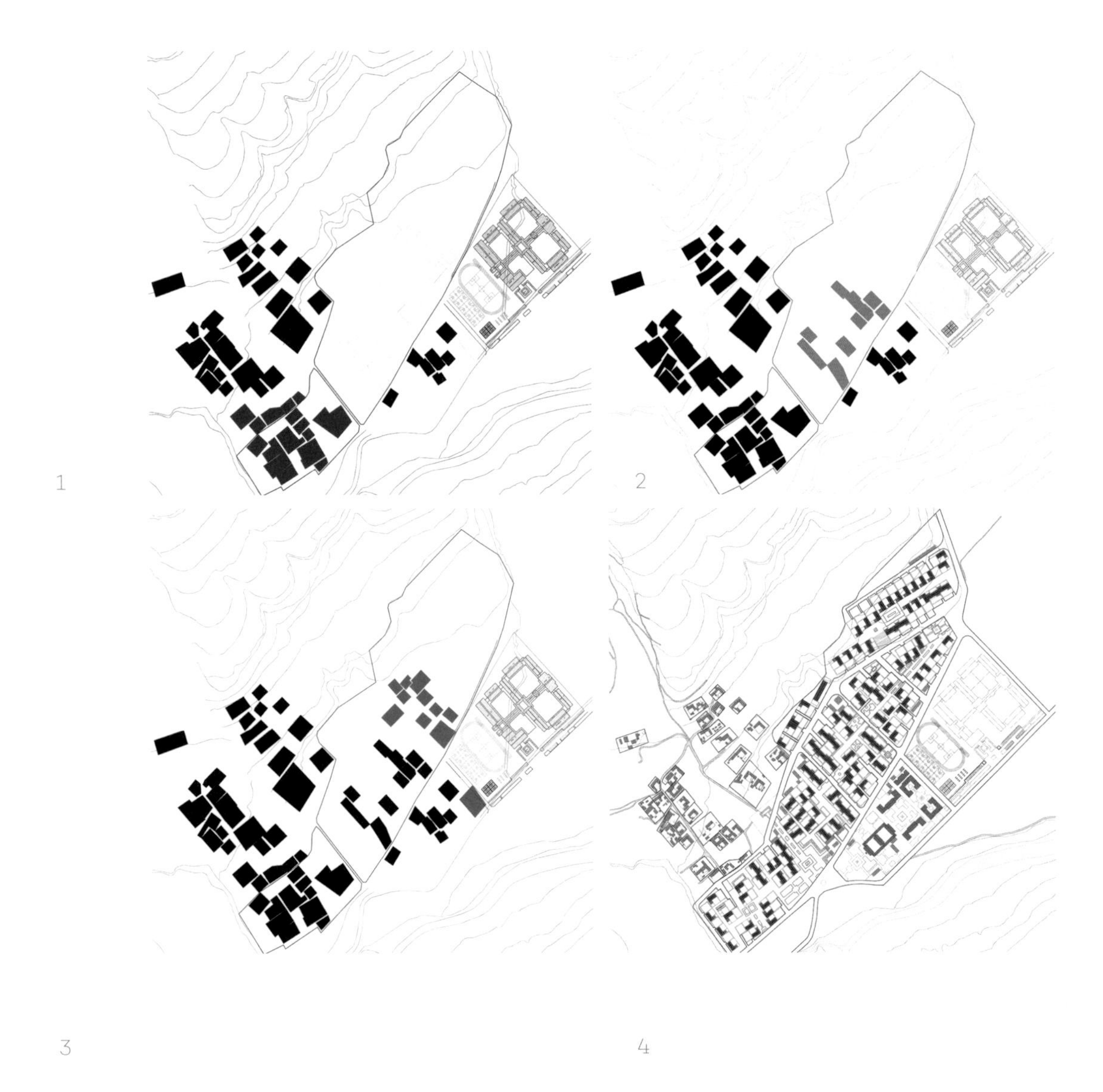

多样化的院落布局，合理对院落的空间属性进行区分

Diversity of Courtyard Layout, Reasonable to Distinguish the Space Properties of the Courtyard

家庄 人口约：200 人
a Jia Zhuang Population: 200

丁家庄 人口约：200 人
Ding jiazhuangZhuang Population: 200

罗景村 人口约：500 人
Lo king Tsuen Population: 500

石家窑坡 人口约：100 人
Stone House Kiln Slope Population: 100

家营村 人口约：160 人
e Jiaying Village Population: 160

寨子村 人口约：80 人
Zhaizi Village Population: 80

郑家沟门 人口约：300 人
Zhengjiagou Gate Population: 300

潘家庄 人口约：100 人
Panjiazhuang Population: 100

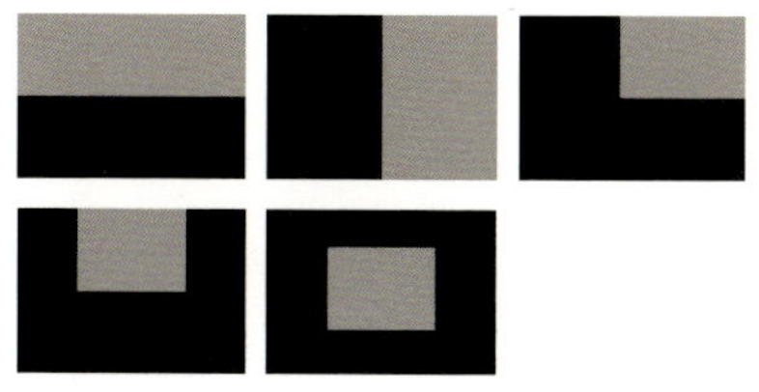

现有村庄院落形态
The Form of the Existing Village Courtyard

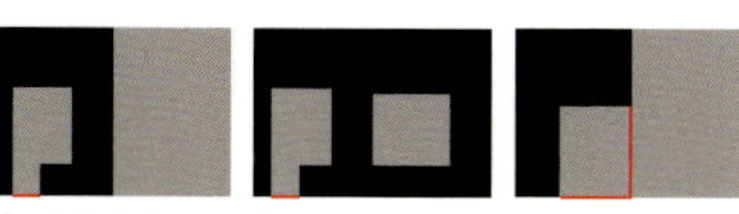

新型北方民居形态
New Form of Northern Dwellings

户型 A：适用于拥有较多农用家具，需要较为独立的家庭
More Agricultural Furniture Independent Family

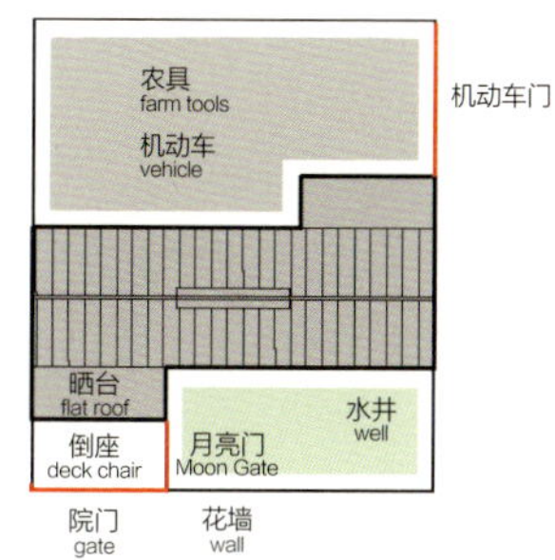

户型 B：适用于家庭农作较少，追求更多南向采光的家庭
Less Family Farming · More South-facing Lighting

建筑
Building

院子
Courtyard

TRADITIONAL
HOTELS
A4

多少日夜奔波坚持
是一份追求理想的执念。
Spending Days and Nights Keeping Running Around Is a
Obsessiveness of Pursuing Dreams.
c3
one
studio

后记 / 造房子是一件幸福的事情

建筑师似跑马

建筑师总处在马不停蹄的状态中。策划、讨论、出差、汇报、修改再轮回。纷繁的工作似乎把时间和思路撕裂成各种不连续的片段，每个项目从拿到任务开始就是“跑马”的时间要求，“跑马”似乎已成为建筑职业的节奏常态，甚至是这个时代的常态。

趋于合理的本能

尽管“时代抛弃你时，连一声再见都不会说”[1]，但我们并不着急随大流“生产”，我们很重视每段设计概念、逻辑、形式、空间之间的合理性贯穿。虽然建筑往往关联庞大资源，我们总或多或少地期望在设计中追求相对庞大的叙事，但在这个世界中，许多逻辑合理小事件的叠加才是生活的本源。而通过恰当的设计来感染建筑背后的生活方式，是我们追寻成就感的真正源泉。

困难大于荣誉

建筑的建成和艺术品诞生的最大不同，便是它不是一场设计师的独舞，建筑师需要统筹各方资源，平衡最佳方案。在与业主、市场、工艺、日常观念的周旋、协调和坚持中，提出最有效的运营策略；基于职业身份的责任感，提出更有利于城市及社会发展的规划策略；以 " 人本 " 为出发点，提出最契合使用需求的设计策略。这些能力的综合需要长期的经验积累，而成一作为一支年轻的团队，因为定位特殊[2]，在面对这些多线程与高挑战的项目时，我们所面临的困难也可想而知。

不完美主义

我们很想把每个作品做到完美，但事实上，在职业生涯中，建筑师不得不面对各种“不完美”和“权宜之计”。伍重在悉尼歌剧院中不得不以混凝土拱来代替薄壳；苏夫洛不得不封堵巴黎先贤祠的侧窗以抵抗穹顶的侧推力；陆谦受不得不更改外滩中国银行大楼的高度以应对建设资金的不足，但这些不完美并不会影响这些作品的魅力和恒久。“不完美”是我们的一种进行时态，也是追求下一个完美作品的态度。

新作品不受表扬的规律

隈研吾说：“熟悉了以前旧建筑的人们，对新建筑绝不会表扬……大家都觉得还是以前的好。”但推动时代建筑向前发展的却往往又是这些创造不同于过往旧建筑的建筑师。没有柯布或密斯，我们或许还在争论如何在高层办公楼中使用柱式；没有巴拉干或文丘里，我们或许还在遵从国际式的法则而忽略地方性的差异。新的作品总会收到“习惯”和“熟悉”的质疑，我们的作品也常因为“不同于以往”而被质疑，但我们永远不会放弃寻找“最合理”。

走在探索的路上的成一工作室

我们坚信设计的价值，不放弃创新和追求卓越的信念。在精耕细作的地产开发如今，市场将重返价值需求的本源。即使如今设计界仍有粗放开发的思维惯性，但有格调的定制化的设计才是发展的潮流。建筑师的职业之路是漫长的，对于刚刚启程的我们，这本作品集是起跑线上的第一个清晰坐标。我们将继续怀着耐心与勇气，走在探索的路上，勤勉而笃实。

这本《成一册》是工作室组建前后在 2014-2017 年部分工作的总结，前后历经四年时间，经历了大约 36 个项目及片段，本册作品集选择了其中 18 个项目，作为系统的总结与回顾。

王涌臣
成一工作室 主持建筑师
2017 年 12 月 上海

注：

[1] 摘自“领航者大会——暨 2018 品牌医生年度盛典”张泉灵演讲内容，背景是阿里巴巴收购大润发，传统王者被新强者替代。瞬息万变的时代，新鲜事物不断涌现，思维方式更迭换代，我们生活的时代充满了巨大的不确定性。

[2] 成一工作室是定位于水石平台的先锋设计团队，在城市更新、产业地产、公共建筑、特色公寓等更多领域发展，开拓并建立新的核心竞争力。

Postscript / Building a House is a Happy Thing

Architects are Like Running Horses

Architects is always in a non-stop state: plan, discussion, travelling, report, modify and re-circulate. The tedious work seems to tear time and ideas into discontinuous segments. Starting from getting the task, each project is the time requirement of "Running Horses". The "Running Horses" seems to have become the rhythm normality of the architectural profession, even the normality of this era.

The Instinct that Tends to be Reasonable

Although "When the times abandoned you, they couldn't even say goodbye", we did not rush to "production" with the big stream. We attached great importance to the rational penetration of concepts, logic, forms, and spaces in each section of the design. Although buildings are often associated with huge resources, we are always more or less expected to pursue relatively large narratives in the design. However, in this world, the superposition of many logical and reasonable events is the source of life. Infecting the lifestyle behind the building through proper design is the real source of our sense of achievement.

Difficulties are More Than the Honor

The biggest difference between building and the birth of art is that it is not a designer's solo, architects need to co-ordinate resources and balance the best solutions. The most effective operation strategy is put forward in the process of dealing with the owner, the market, the craft, the daily idea, and the most efficient operational tactics, based on the sense of responsibility of the professional identity, proposing the planning strategy which is more favorable to the development of city and society; based on "Humanism", the design strategy The combination of these capabilities requires long-term experience accumulation, and as a young team, because positioning Special 2, in the face of these multi-threaded and high challenges of the project, we are faced with the difficulties can be imagined.

Imperfectism

We really want to make every work perfect. But in fact, in the professional life, the architect has to face all kinds of "improper" and "expedient measures." In the Sydney Opera House, Wu Zhong had to replace the thin shell with a concrete arch. Suvlos had to seal the side window of the Parisian Pantheon to resist the side thrust of the dome. Lu Qian had to change the height of the building of China Bank to deal with the lack of construction funds. But these imperfections do not affect the charm and longevity of these works. "No perfection" is a kind of continuous tense of us. It is also the attitude of pursuing the next perfect work.

The Regularity that are Not Subject to be Praised of New Works

Kengo Kuma said: "People who are familiar with the old buildings will never praise the new building... everyone thinks it was good before." But it is often the architects who create different buildings from past old buildings. Without Cobb or Mies, we may still be debating how to use columns in high-rise office buildings. Without Balacan or Venturi, we may still follow international rules and ignore local differences. New works are always questioned by "custom" and "familiarity". Our works are often questioned because they are "different from the past." But we will never give up searching for "the most reasonable thing".

One Studio Studio on the Road of Exploration

We firmly believe in the value of design, and do not give up the faith of innovation and the pursuit of excellence. In the intensive development of farming, the market will now return to the origin of value needs. Even though the community of design still has the extensive inertia of thought, the custom design with style is the trend of development. The career path of the architect is long. For us who have just set off, this sample reels is the first clear coordinate of the starting point. We will continue to walk on the road of exploration with patience and courage and be diligent and conscientious.

This book is a summary of part of the work before and after the establishment of the studio in 2014-2017. It went through about four years and experienced about 36 projects and clips. This booklet selected 18 items as a systematical summary and review.

Yongchen Wang
One studio Presiding architect
Shanghai, Dec. 2017

Note:

1 It is extracted from the content of Zhang Quanling speech of the "Leadership Conference - cum 2018 brand doctor annual festival". The background is Alibaba's acquisition of RT-Mart and the traditional king was replaced by the new strong. In the ever-changing era, new things continue to emerge. Patterns of thinking patterns changes. Our era of life is full of great uncertainty.
2 One studio is a pioneering design team positioned on the W&R platform. It develops in areas such as urban renewal, industrial real estate, public buildings, and specialty apartments, and explores and establishes new core competitiveness.

建筑师的黄金时代

中国已连续近十年成为世界上 200 米以上高层建筑竣工面积最多的国家，伴随着每年新增建筑面积的迅猛增长，亦成为全球每年完工面积第一的建筑大国。对于起步较晚的国内建筑设计行业来说，这正是中国建筑师的黄金时代，在国家高速城市化的节奏之下，有着大量的项目实践机会，设计和创意以飞快的速度得到实现，中国建筑师同期的经验积累以及专业成长速度甚至超过了西方。对于处在发展浪潮之中的建筑师来说，这是完成行业经验原始积累的宝贵机遇。虽然在公共建筑的设计领域中仍然有比较高的粗放性和不确定性，但我们始终确信，唯有从实际项目中的点滴积累，才能得到最宝贵的经验和历练，这是时代的契机，也是成为一名成熟建筑师的必经之路。

设计首先是解决问题

把建筑师想象成严阵以待的艺术家英雄，为创造出最伟大的文明成就而努力不懈，这种浪漫的想法曾经最早由歌德等作家们率先提出。如今建筑师的角色定义已经大大拓展，除了掌握基本的设计原则和方法，更要对社会发展和生活方式的变革等方面有独立的思考。抛开诸多的理论学说去思考本质，设计首先是解决问题。不论是人和建筑的问题，或是建筑和城市的关系，还是功能和形式之间的矛盾，通过恰当的设计来统筹协调项目中的各类资源，是建筑师最大的价值所在。如今社会的发展日新月异，新的理念和技术不断涌现，新的设计工具也非常先进，但分析和解决问题的原理并没有改变，这是我们建筑设计创作的根基。

为最终使用者做设计

设计为谁而做，这是个必须明确但往往又被忽视的问题。作为服务行业的准则之一，满足客户的要求已经成为了政治正确，设计师似乎唯有努力迎合。相对于医院或学校这类使用对象相对明确的建筑类型，更多的建筑类型使用者是更为广泛的人群。对不同人群特性的理解往往会体现出很大的差异性，设计决策者很容易根据个人的喜好而把自己变成使用者的化身，对设计师提出各种不合理的要求。设计师必须时刻保持头脑清醒，以真实客体的使用体验为中心，合理的引导设计回归到正轨上来。能否基于最终使用者的真实需求做设计，隐含着建筑设计的基本价值观，也是设计师对自己职业意义的理解和选择。

追求设计和建造的统一

优秀的设计一定可以构建一个精彩的虚拟世界，但最终效果的实际呈现才是建筑设计的向社会大众的最终传达。设计和建造密不可分的传统由来已久，在中世纪，建筑的设计和施工程序都是在石匠师傅的指挥下完成的，他们必须精通几何学，不但要能勾勒出复杂的石头造型，还要能为整体建筑提供统一的空间和结构形式，最终需要经过一整套工匠学徒训练才能培养出来。如今的建筑师仍然肩负着将设计最终落地实现的使命，从材料的选择到细部的控制，无一不体现出了艺术性和技术型的高度结合。在项目落地的过程中提高设计还原度，追求设计和建造的高度统一是我们不懈的追求

时间是设计的最终检验者

2018 年，美国建筑师学会（ American Institute of Architects ）发布新年公告，“二十五年奖”的评审团没有找到在这二十五年间不仅在美学与文化影响上做出突出贡献，而且还对建筑行业产生影响的不朽的建筑，本年度“二十五年奖”将取消。这是自 1971 年该奖项成立近五十年以来，第一次发生这种状况。这种认真甚至刻板的态度非常值得敬佩，提醒着我们慎重的对待设计中的每一个决定，并思考什么样的建筑才能获得“经得起时间的考验”这样极高的评价。“二十五年奖”代表着一种永恒的场所精神，设计成为了穿越时空的印记。在宏观的时间轴上，实现长久的设计品质更是设计的意义所在，也是我们长期的追求。

走在探索的路上

建筑师的职业之路是漫长的，对于刚刚启程的我们，这本作品集是起跑线上留下的第一个印记。我们仍将坚信设计的价值，不放弃创新和追求卓越的信念。在精耕细作的地产开发节奏下，在顺应市场的激烈竞争中，设计自有千钧之力。我们也将继续怀着耐心与勇气，走在探索的路上，勤勉而笃实。

这本《成一册》是工作室组建前后在 2014-2017 年部分工作的总结，前后历经四年时间，经历了大约 36 个项目及片段，本册作品集共选择了其中 18 个项目，作为系统的总结与回顾，不足之处，还望指正

赵延伟
成一工作室 主持建筑师
2017 年 12 月 上海

The Golden Age of Architects

For nearly ten years, China has become the country with the largest number of high-rise buildings over 200 meters in the world. With rapid growth of new buildings, it has also become the country with the first completion area annually in the world. For the late domestic architectural design industry, it' s the golden era of Chinese architects. With rhythm of high-speed urbanization, there are many opportunities for project practice. Design and creativity are realized at a rapid pace. Chinese architects' speed of experience accumulation and professional growth over the same period even surpassed that of the West. For architects in the wave of development, it' s a precious opportunity to complete original accumulation of industry experience. Although there is still a relatively high degree of extensiveness and uncertainty in the design of public buildings, we believe that only the slightest accumulation in practical projects can yield the most valuable experience, which is an opportunity for the times and the only way to become a mature architect.

Design is First to Solve the Problem.

Imagining architects as heroes of artists who work hard to create the greatest achievements of civilization. The romantic idea was first proposed by Goethe and other writers. Today, the definition of architects is greatly expanded. Besides basic design principles and methods, it' s necessary to have independent thinking on social development and lifestyle changes. Aside from many theoretical doctrines to think essence, design is first to solve the problem. Whether it' s the problem of people and buildings, or the relationship between architectures and cities, or the contradiction between function and form, it is architects' greatest value to coordinate various resources through proper design. Nowadays, with rapid development of society, new ideas and technologies are emerging and new design tools are also very advanced. But principles of analyzing and solving problems don' t changed, which is the foundation of our architectural design.

Design for Ultimate Users

Designing for whom is a question that must be clear but often overlooked. As one of the guidelines for service industry, satisfying customer requirements become politically correct. Designers seem to work hard to cater. For relatively clear types of buildings such as hospitals or schools, more users are more extensive. Understanding of characteristics of different groups often reveals great differences. Design decision makers can easily turn themselves into incarnations of users based on personal preferences and make various unreasonable demands on designers. Designers must always keep their heads clear, focus on the use of real objects, and guide the design back to the right track. Whether the design can be based on real needs of ultimate users implies basic values of architectural design, which is also designer's understanding and choice of their own professional meaning.

Pursuing the Unification of Design and Construction

Excellent design can certainly build a wonderful virtual world. But actual presentation of the final result is the ultimate communication of architectural design to the public. The tradition that design and construction are inextricable is long-standing. In the Middle Age, architectural design and construction procedures were completed by command of masonry masters. They must be proficient in geometry, sketching complex stone shapes and providing a unified space and structural form for the entire building. It ultimately needs a set of artisan apprenticeship training. Today's architects are still shouldering the mission to achieve the final design. From choice of materials to control of details, they reflect high-degree artistic and technical integration. It is our unremitting pursuit to increase the degree of design reduction and pursue high-degree unity in design and construction in the process of landing the project.

Time is the Ultimate Checker of Designing.

American Institute of Architects issued a New Year' s announcement in 2018. The judging panel of the "Twenty-five-year Award" didn' t find that during these twenty-five years, it not only made outstanding contributions to aesthetic and cultural influences, but also immortal buildings that have an impact on the construction industry will be cancelled this year. It' s the first time this kind of situation has occurred since the award was established in 1971. This serious and even rigid attitude is very worthy of admiration, reminding us to treat every decision prudently and thinking what kind of building can get such a high rating of "the test of time". "Twenty-five Year Award" represents an eternal place spirit designed to become a mark through time and space. On the macro timeline, the realization of long-term design quality is the meaning of design and our long-term pursuit.

Walking on the Road to Exploration.

The career path of the architect is long. For us who just started, the portfolio is the first mark left on the starting line. We firmly believe in the value of design and don' t give up the faith of innovation and the pursuit of excellence. Under rhythm of intensive development of real estate, in the fierce competition in response to market, design has its own strength. We will also continue to work with patience and courage on the road to exploration and be diligent and conscientious.

This book is a summary of part of the work before and after the establishment of the studio from 2014 to 2017. After four years, it has undergone about 36 projects and clips. In the booklet, 18 projects were selected as a systematic summary and review. Please point out and correct the insufficiency.

Yanwei Zhao
One studio Presiding architect
Shanghai, Dec. 2017

2015.4 青岛 . 中联产业园
出乎意料的驻场几日
Unexpected Stay for a Few Days

2015.9 深圳 . 世茂五象国际中心
关键节点汇报日，下午两点的午餐
Key Node Reporting Day, Lunch at 2:00pm

2015.10 兰州 . 兰州文化产业中心
次晨汇报，星夜必须赶达
Drive Through the Starry Night
for the Next Day Prestation

2015.12 上海 .《城市再生中的开发与设计》
伴随出版，鲜为人知的十日蹲守
Ten Days and Nights of Hardworking
for Book Publishing

2015.12 北京 . 龙湖门头沟新城
航空管制，遭遇八小时候机
Air Control, Waiting for 8 Hours

2016.3 西安 . 金泰湾盛泰澜度假酒店
首次酒管接洽，被质疑的压力
Being Questioned in the First Time of
Hotel Management

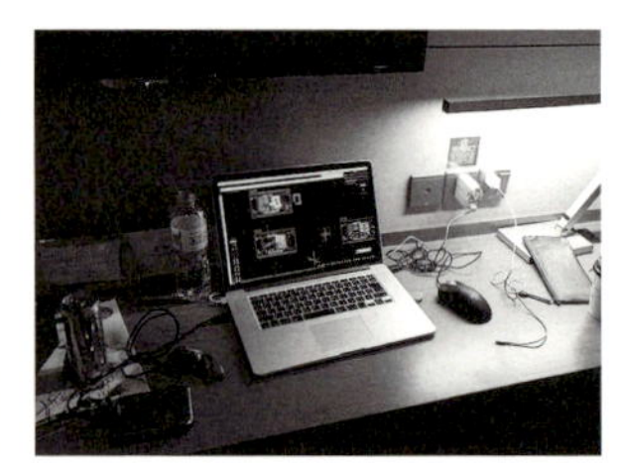

2016.3 郑州 . 和昌航空港区医药园区
凌晨到达酒店，两小时方案梳理
Two Hours of the Project Pectination
before the Dawn

2016.7 海口 . 中海中信广场
三小时，快速思路整理汇报
Three Hours Ideas Arrangement Prestation

2016.7 上海 . 绿地创新产业中心
台风造访背景下的反复修改
Repeated Revisions
in the Typhoon

2017.7 南京 . 瑞安南京珠江路未来城
三班倒的现场与设计抢工
Three Shifts of the Construction
and Design Work

2017.7 上海 . 光明垂直森邻
各路高手的等待汇报时刻
Waiting for Prestation

2017.12 长春 . 水文化园第七净水车间
大雪后的工地拜访
Site Visit After Heavy Snowfall

2015.12 上海，纽斯出汗记
Steaming in NEWS

2015.12 红坊，曾经的 C3 战队
C3 Working Group at Red Town

2016.5 淀山湖，大厨聚餐记
Team Party, Dianshan Lake

2016.7 红坊捕蝉记
Catching in the Red Town

2016.7 头脑风暴轮回记
Brainstorming

2016.11 伟哥庆生记
Birthday Celebration

2017.1 深冬剥蟹记
Crabs Party in the Deep Winter

2017.5 “珠管办”庆功宴
Celebration Feast

2017.12《成一册》诞生讨论记
Book Discussion

2014.6 深圳 . 汇报后看京基 100
Viewing Kingkey 100 after Prestation

2014.6 兰州 . 山路
Lanzhou, Mountain Road

2015.4 东京 . 天光
Tokyo , Sky Light

2015.12 嘉峪关 . 雪山与高铁
Jiayuguan, Snow Mountain and High-speed Rail

2016.7 海口 . 黄昏海景
Haikou, Dusk Seascape

2016.7 南宁 . 有层次的云
Nanning, Hierarchical Cloud

2016.7 三亚 . 迎风出发
Sanya, Starting in the Wind

2016.9 东方 . 夕阳海景
East, Sunset Seascape

2016.12 上海 . 回家
Shanghai, Back Home

2017.7 青岛 . 向黎明出发
Qingdao, Depart in the Dawn

2017.12 西安 . 黄土峡谷
Xi'an, Loess Canyon

2017.12 哈尔滨 . 在云端
Harbin, In the Clouds

致谢

外界的声音很多，灰色地带无处不在，很多莫名其妙的感受、感情，很难说得清楚。当然，有时我们也不是特别坚定自己的想法。但完成这本作品集后，我们更坚定了一些。认真奋斗过的青春是最美好的，我们告诉自己，无论外界多么浮躁和急功近利，永远不要违背自己的内心。

这本作品集里蕴含的精神，对于那些带着懵懂的激情，想要努力过得精彩纷呈的朋友们，是有帮助的。不管环境如何残酷，我们需要珍视内心的柔软，保持着对这个世界的感动。

感谢成一工作室的所有成员和参与过每一个项目的所有人所付出的努力，感谢每一个曾经支持我们工作并尊重设计的业主，感谢所有关注、关心我们工作室成长的朋友。祝福身边的每一个人，愿你们的人生满载最美好的心灵。

《成一册》，制作完结。
2018 年 3 月 15 日

Acknowledgement

"There are many voices from the outside world. Gray areas are everywhere. Many inexplicable feelings and emotions are difficult to say clearly. Of course, sometimes we are not particularly firm with our own ideas. But after finishing the portfolio, we are more firm. The youth who have struggled is the best. We tell ourselves that no matter how impetuous the outside world is and eager for quick success, we must never run counter to our own innermost feelings."

"The spirit contained in the portfolio is helpful for those friends who have a passion for ignorance and want to work hard to have a brilliant life. No matter how cruel the environment is, we need to cherish the softness of our hearts and maintain affection for the world."

We thank all members of One Studio and all the people involved in each project for their hard work. We thank all the owners who once supported our work and respected the design. We thank all friends who care and care about the growth of the studio. Bless everyone around you and wish your life is full of the best of hearts.

One Studio Book, Finished
March 15, 2018

Special thanks

Zhujingye 朱敬业
Leiyu 雷宇
Sunbing 孙兵
Yupeiqiong 余佩琼
Liwei 黎伟
Lianzhiyang 连志阳
Yongweiwei 雍为为
Dulei 杜磊
Qiaoguichun 乔桂春
Lihuizhen 李慧珍
Sunle 孙乐
Qiweijuan 戚卫娟
Wangshen 王申
Lijing 李倞
Chenyuanjian 陈元监
Xudongbo 徐东波
Denggang 邓刚
Duzhiyang 杜治洋
Hezhendong 贺镇东
Chengxiaoyan 程晓燕
Niliang 倪量
Lishougen 李寿根
Suyifan 苏逸凡
Xuyidan 徐艺丹
Tongchenchen 童晨晨
Yanzhi 严志
Shiyili 施益丽

NEVEREND

Zhourui 周睿
Lianglihua 梁莉华
Wangxuan 王煊
Xiepeng 解澎
Huzhuolin 胡卓霖
Zhenglibin 郑丽彬
Shenhe 沈禾
Zouleilei 邹蕾蕾
Yuanxiaozhong 袁小忠
Lilan 李岚
Zhouyi 周怡
Dengsong 邓松
Caoye 曹野
Yupeiqin 余佩钦
Yanghao 杨浩
Xialili 夏丽丽
Zhoutiegang 周铁刚

REVOLUTION

Wangjin 王进
Gongbing 龚兵
Huoyuanlong 火元龙
Lixuting 李旭庭
Zhulei 朱蕾
Zhengjian 郑健
Lirui 李瑞
Liutiezhu 刘铁柱
Wangjie 王杰
Wuyin 吴寅
Shenhongze 沈宏泽
Wangsuowen 王锁文
Zhenyongqi 郑永琪
Jiazhonghu 贾仲瑚
Wujiyu吴计瑜

超高层办公、主题公园及城市更新

SUPER HIGH-RISE OFFICE, ThEMATIC PARK & UPDATING of the CITY 2014-2017

项目名称 Project Name	面积 Area	类型 Type	状态 Situation	年份 Years
兰州文化产业发展孵化中心 Cultural Industry Development Incubation Center	273 720 m²	办公 / 超高层 Office/Super High level	方案 Programme	2014
兰州南关十字润通广场 South Gate Runtong Square	229 700 m²	办公 / 超高层 Office/Super High level	方案 Programme	2015
上海绿地创新产业中心 Greenland Innovation Industry Center	62 200 m²	产业园 / 总部办公 Industrial Park/HQ Office	建成 Completed	2014
上海国际能源创新中心 International Energy Innovation Center	897 100 m²	产业园 / 城市更新 Industrial Park/City Update	在建 Building	2015
南京珠江路创客大街城市更新 Zhujiang Road Chongke Street City Update	37 900 m²	办公 / 城市更新 Office/City Update	在建 Building	2017
上海杨浦区光明集团申宏冷库改建 Shenhong Cold Storage Reconstruction	78 750 m²	产业园 / 城市更新 Industrial Park/City Update	在建 Building	2016
上海南翔小绵羊总部产业园 Industrial Park Headquarters of NanxiangSmall Sheep	58 864 m²	产业园 / 城市更新 Industrial Park/City Update	方案 Programme	2017
上海嘉定南翔游戏主题产业园 Nanxiang Game Theme Industrial Park	76 705 m²	产业园 / 城市更新 Industrial Park/City Renewal	方案 Programme	2017
上海青浦区海博西虹桥物流产业园 Haibo West Hongqiao Logistics Industrial Park	89 900 m²	产业园 / 专业园区 Industrial Park/Professional Park	在建 Programme	2017
郑州航空港区生物医药产业园 Aviation Port Biomedical Industrial Park	504 815 m²	产业园 / 专业园区 Industrial Park/Professional Park	方案 Programme	2017
上海赤峰路光明进修学校 Bright Training School of Chifeng Road	9 160 m²	产业园 / 城市更新 Industrial Park/City Update	方案 Programme	2017
长春净水厂城市再生第七净水车间 City Regeneration of Water Purification Plant	701 m²	产业园 / 城市更新 Industrial Park/City Update	建成 Building	2017

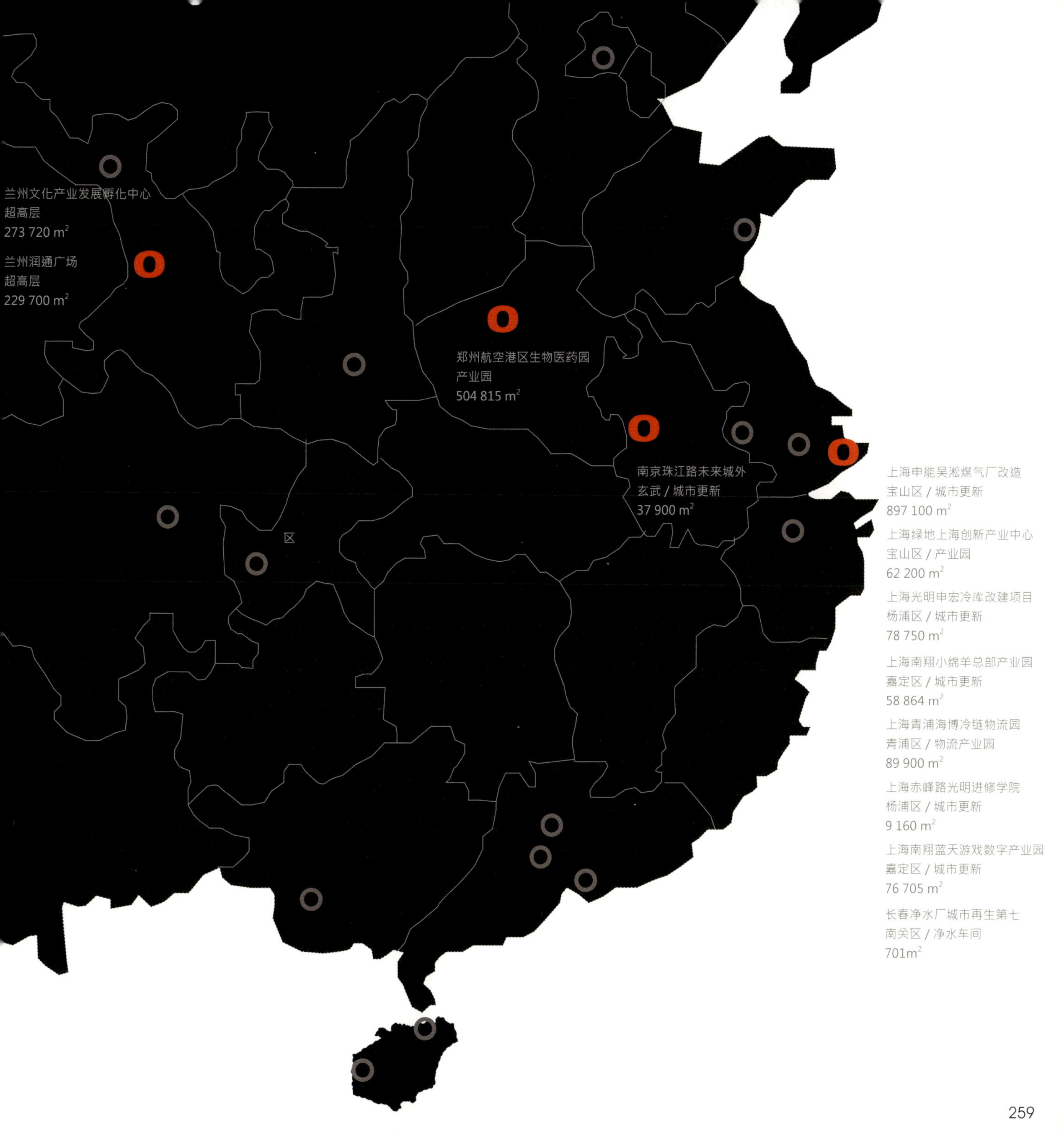
兰州文化产业发展孵化中心
超高层
273 720 m²
兰州润通广场
超高层
229 700 m²
郑州航空港区生物医药园
产业园
504 815 m²
南京珠江路未来城外
玄武 / 城市更新
37 900 m²
上海申能吴淞煤气厂改造
宝山区 / 城市更新
897 100 m²
上海绿地上海创新产业中心
宝山区 / 产业园
62 200 m²
上海光明申宏冷库改建项目
杨浦区 / 城市更新
78 750 m²
上海南翔小绵羊总部产业园
嘉定区 / 城市更新
58 864 m²
上海青浦海博冷链物流园
青浦区 / 物流产业园
89 900 m²
上海赤峰路光明进修学院
杨浦区 / 城市更新
9 160 m²
上海南翔蓝天游戏数字产业园
嘉定区 / 城市更新
76 705 m²
长春净水厂城市再生第七
南关区 / 净水车间
701m²
区

商业、酒店及公共建筑

COMMERCE,HOTEL & PUBLIC ARCHITCTURES 2014-2017

项目名称 Project Name	面积 Area	类型 Type	类型 Situation	年份 Years
南宁世茂五象国际中心 Shimao Five Elephant International Center	325 263 m²	超高层 商业 Super High-rise Business	在建 Construction	2014
上海崇明长岛游艇码头区 Chongming Long Island Yacht Wharf Area	76 255 m²	城市设计 Urban Design	方案 Programme	2015
广州实地集团禾丰 9# 地块 No. 9th Block, Wo Fung, Guangzhou Field Group	108 640 m²	街区商业 Block Business	方案 Programme	2014
嘉峪关东湖商业步行街 Dong Hu Commercial Pedestrian Street	40 456 m²	商街改造 Commercial Street Renovation	方案 Programme	2014
嘉凯城袍江城市客厅 Jakarta City Porsche City Living Room	30 556 m²	街区商业 Block Business	方案 Programme	2016
苏州金鸡湖 Indigo 酒店 Jinji Lake Indigo Hotel	75 201 m²	五星级酒店 Five-star Hotel	方案 Programme	2016
海南盛泰乐金泰湾度假酒店 Shengtai Le Jin Taiwan Resort Hotel	115 967 m²	五星级酒店 Five-star Hotel	在建 Construction	2017
兰州第一人民医院业务综合楼 Lanzhou First People 's Hospital Business Complex	45 682 m²	医院 Hospital	在建 Construction	2016
深圳金地未来系产品研发 Shenzhen Gemdale future series products research		研发 Research	方案 Programme	2014
兰州城关区两场一馆 Lanzhou Public Complex in Chengguan District	45 682 m²	公共建筑 Pubilc Buildings	方案 Programme	2015
成都龙湖时代天街策划 Chengdu Longhu Age Yin planning	651 094 m²	策划 Planning	方案 Programme	2015
兰州榆中县沿川文园 Lanzhou Yanchuan Lake Garden in Yuzhong country	30 600 m²	村落 / 城市更新 Village/City Update	方案 Programme	2016
青岛金地世家启动区 Qingdao Gemdale Aristo Promoter District	2 361 m²	精品住宅 Quality Residential	方案 Programme	2017

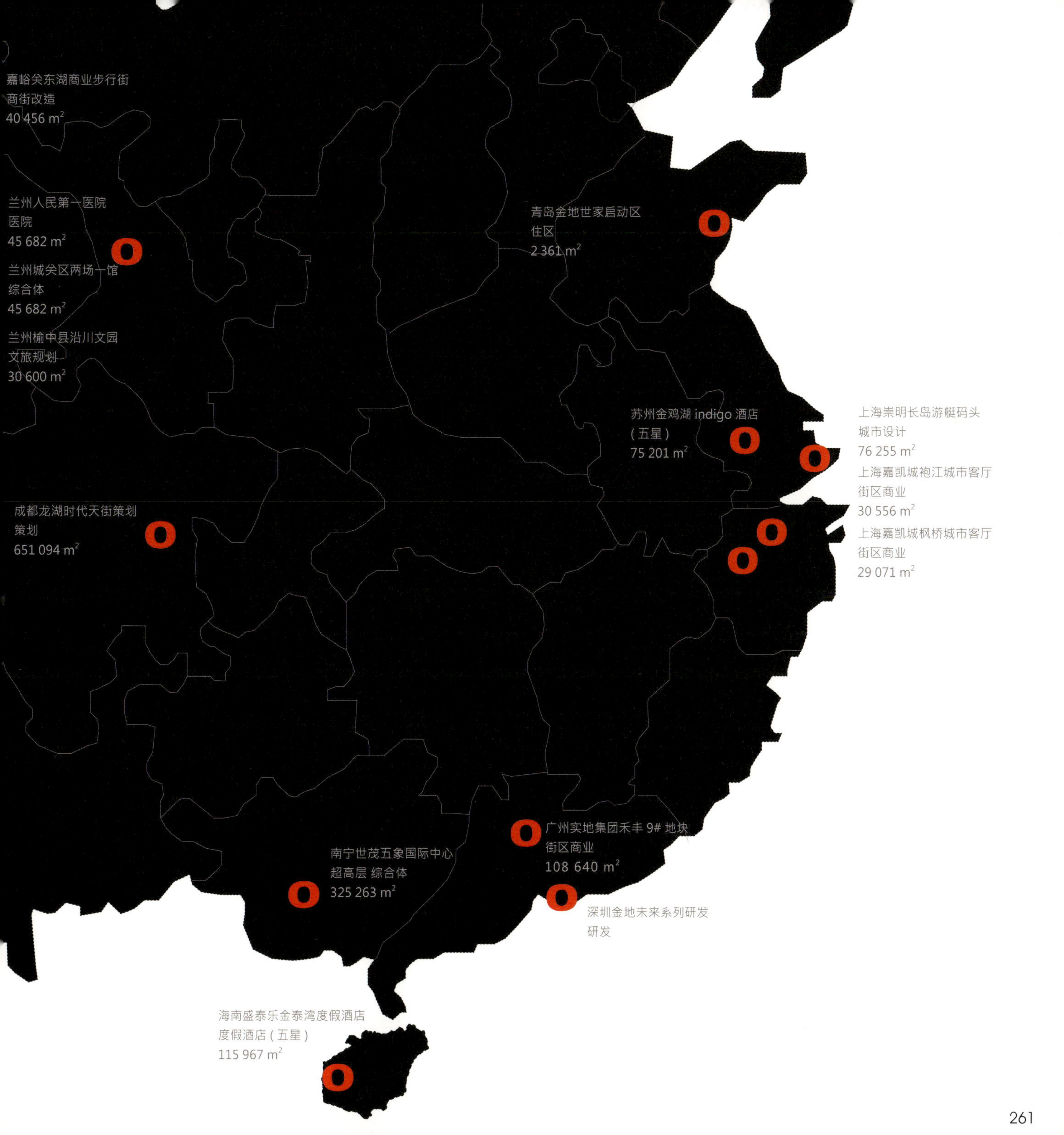

嘉峪关东湖商业步行街
商街改造
40 456 m²
兰州人民第一医院
医院
45 682 m²
兰州城关区两场一馆
综合体
45 682 m²
兰州榆中县沿川文园
文旅规划
30 600 m²
青岛金地世家启动区
住区
2 361 m²
苏州金鸡湖 indigo 酒店
（五星）
75 201 m²
上海崇明长岛游艇码头
城市设计
76 255 m²
上海嘉凯城袍江城市客厅
街区商业
30 556 m²
上海嘉凯城枫桥城市客厅
街区商业
29 071 m²
成都龙湖时代天街策划
策划
651 094 m²
广州实地集团禾丰 9# 地块
街区商业
108 640 m²
南宁世茂五象国际中心
超高层 综合体
325 263 m²
深圳金地未来系列研发
研发
海南盛泰乐金泰湾度假酒店
度假酒店（五星）
115 967 m²

水石设计
W&R Group

"水石设计"成立于1999年，致力于打造专业化和规模化的设计平台。旗下包括上海水石建筑规划设计股份有限公司（证券代码871658）、上海水石景观环境设计有限公司、上海水石城市规划设计有限公司、上海水石工程设计有限公司等多家设计机构。总部位于上海，在重庆、深圳设有分公司，并在沈阳、青岛、西安、武汉、苏州、成都、南宁等设有办事处，员工人数达1000人。

水石设计拥有建筑行业建筑工程甲级资质、风景园林工程设计专项甲级资质、城市规划编制乙级资质，其核心技术能力涉及精品住宅、商业综合体、主题产业园、城市再生等类型；服务范围涵盖规划设计、建筑设计、景观设计及室内设计等多专业和全过程的设计及咨询服务。

水石设计倡导精细化和一体化设计，强调价值挖掘、产品创新、品质稳定度，以及设计图纸与建成效果的高还原度。长期服务于众多知名开发商，包括金地、龙湖、保利、万科、绿地、中南、世茂、金茂、泰禾、中海、华润、华发等，目前建成作品覆盖全国150多个城市。

相关出版物
Relevant Publications

项目类	“绿地长春饕界” “绿地闸北中央广场” “样板区一体化设计” “石库门复兴——90墅系列产品之‘公元1860’”
专题类	水石国际、精品住宅、城市社区商业、城市再生、景观设计、耇 水石十年
正式出版物	《城市主题产业园设计与开发》、《地产模式下的精细化设计》、《耇·建文筑章》 《城市公园设计》、《城市公园设计》（教材）、《城市再生中的开发与设计》

行业影响力
The Influence of the Vocation

2017年 水石挂牌新三板，证券名称“水石设计”，证券代码871658

2017年 水石成功中标长春水文化生态园EPC设计施工一体化项目

2017年 水石改造并入驻DESIGN188设计园区

2016年 水石受邀担任高校国际联合教学答辩“帝国的边界：罗马terminni火车站周边地块城市更新”特邀评委

2016年 水石与红坊共同举办“2016城市再生论坛——探索2020年红坊的形态与业态”

2015年 水石主办“城市再生价值谈”论坛

2015年 水石成立“水石国际城市再生设计中心”

2015年 水石就“旧工业遗产艺术改造”话题接受湖南卫视采访

2015年 水石应邀出席西安老钢厂设计创意产业园举办的“城市·地产·设计”思想创新总裁峰会

2014年 水石主办M7当代中国青年建筑师主题沙龙

2014年 水石应华东院邀请参与“上海市建筑业志设计篇”关于历史保护建筑编撰工作

2011年 水石主办“无缝对接——城市主题产业园发展”论坛

2007年 水石协办深圳香港双城双年展的分会场“城市再生论坛”

one studio

成一工作室

Insightful Architecture

An Insightful Design Studio
Since a.d. 2017, by
W&R Group, Shanghai

188 DESIGN
No.188 Guyi Road, Xuhui District,
Shanghai, 200235
+86-021-54679918

出品：
水石成一工作室

编著：
王涌臣

策划：
王涌臣、赵延伟、张妍钰

技术支持：
胡超、王臣、卫琨、谢梅美、于洪浩、段嫣然、
金鹏、沈迎光、全轩震

插画：
张佾媛

装帧设计：
王涌臣

图书在版编目（CIP）数据

One book 成一册 / 王涌臣编著 . -- 上海 : 同济大学出版社 , 2018.4
ISBN 978-7-5608-7807-2

Ⅰ . ①O… Ⅱ . ①王… Ⅲ . ①建筑设计 - 作品集 - 中国 - 现代
Ⅳ . ① TU206

中国版本图书馆 CIP 数据核字 (2018) 第 075958 号

书名：成一册
编著：王涌臣
责任编辑：荆华　　责任校对：徐春莲　　封面设计：王涌臣
英文翻译：胡宸

出版发行：同济大学出版社　www.tongjipress.com.cn
（上海市四平路 1239 号 邮编 200092 电话 021-65985622）
经　销：全国各地新华书店
印　刷：上海雅昌艺术印刷有限公司
开　本：889mm × 1194mm 1/20
印　张：13
印　数：1—1 000
字　数：406 000
版　次：2018 年 4 月 第 1 版　　2018 年 4 月 第 1 次印刷
书　号：ISBN 978-7-5608-7807-2
定　价：188.00 元